CSET 122, 126

Earth and Planetary Science
Teacher Certification Exam

By: Sharon Wynne, M.S.
Southern Connecticut State University

"And, while there's no reason yet to panic, I think it's only prudent that we make preparations to panic."

XAMonline, INC.
Boston

Copyright © 2008 XAMonline, Inc.
All rights reserved. No part of the material protected by this copyright notice may be reproduced or utilized in any form or by any means, electronic or mechanical, including photocopying, recording or by any information storage and retrievable system, without written permission from the copyright holder.

To obtain permission(s) to use the material from this work for any purpose including workshops or seminars, please submit a written request to:

> XAMonline, Inc.
> 21 Orient Ave.
> Melrose, MA 02176
> Toll Free 1-800-509-4128
> Email: info@xamonline.com
> Web www.xamonline.com
> Fax: 1-781-662-9268

Library of Congress Cataloging-in-Publication Data

Wynne, Sharon A.
    Earth and Planetary Science 122, 126: Teacher Certification / Sharon A. Wynne. - 2nd ed. ISBN 978-1-58197-399-0
    1. Earth and Planetary Science 122, 126   2. Study Guides.   3. CSET
    4. Teachers' Certification & Licensure.   5. Careers

Disclaimer:
The opinions expressed in this publication are the sole works of XAMonline and were created independently from the National Education Association, Educational Testing Service, or any State Department of Education, National Evaluation Systems or other testing affiliates.

Between the time of publication and printing, state specific standards as well as testing formats and website information may change that is not included in part or in whole within this product. Sample test questions are developed by XAMonline and reflect similar content as on real tests; however, they are not former tests. XAMonline assembles content that aligns with state standards but makes no claims nor guarantees teacher candidates a passing score. Numerical scores are determined by testing companies such as NES or ETS and then are compared with individual state standards. A passing score varies from state to state.

Printed in the United States of America

CSET: Earth and Planetary Science 122, 126
ISBN: 978-1-58197-399-0

TEACHER CERTIFICATION STUDY GUIDE

Table of Contents

Pg.

Part I: Content Domains for Subject Matter Understanding and Skill in Earth and Planetary Science

Competency 1.0	Understand galaxies and stars	1
Competency 2.0	Understand solar systems	9
Competency 3.0	Understand planets and satellites	13
Competency 4.0	Understand tectonic processes	18
Competency 5.0	Understand oceans	26
Competency 6.0	Understand the atmosphere	34
Competency 7.0	Understand Earth's energy budget including inflow and outflow	38
Competency 8.0	Understand circulation in the oceans and atmosphere	43
Competency 9.0	Understand climate variations in time and space	49
Competency 10.0	Understand the rock cycle	52
Competency 11.0	Understand the water, carbon, and nitrogen cycles	56
Competency 12.0	Understand tectonic evolution	63
Competency 13.0	Identify major economic Earth resources	66
Competency 14.0	Explain surface processes	70
Competency 15.0	Understand natural hazards	76
Competency 16.0	Understand geologic mapping	78

EARTH & PLANETARY SCIENCE

TEACHER CERTIFICATION STUDY GUIDE

Part II: Subject Matter Skills and Abilities Applicable to the Content Domains in Science

Domain 1. Investigation and Experimentation 81

Domain 2. Nature of Science 96

Domain 3. Science and Society 104

Sample Test 113

Answer Key 124

Rationales with Sample Questions 125

Great Study and Testing Tips!

What to study in order to prepare for the subject assessments is the focus of this study guide but equally important is *how* you study.

You can increase your chances of truly mastering the information by taking some simple, but effective steps.

Study Tips:

1. Some foods aid the learning process. Foods such as milk, nuts, seeds, rice, and oats help your study efforts by releasing natural memory enhancers called CCKs (*cholecystokinin*) composed of *tryptophan*, *choline*, and *phenylalanine*. All of these chemicals enhance the neurotransmitters associated with memory. Before studying, try a light, protein-rich meal of eggs, turkey, and fish. All of these foods release the memory enhancing chemicals. The better the connections, the more you comprehend.

Likewise, before you take a test, stick to a light snack of energy boosting and relaxing foods. A glass of milk, a piece of fruit, or some peanuts all release various memory-boosting chemicals and help you to relax and focus on the subject at hand.

2. Learn to take great notes. A by-product of our modern culture is that we have grown accustomed to getting our information in short doses (i.e. TV news sound bites or USA Today style newspaper articles.)

Consequently, we've subconsciously trained ourselves to assimilate information better in neat little packages. If your notes are scrawled all over the paper, it fragments the flow of the information. Strive for clarity. Newspapers use a standard format to achieve clarity. Your notes can be much clearer through use of proper formatting. A very effective format is called the *"Cornell Method."*

> Take a sheet of loose-leaf lined notebook paper and draw a line all the way down the paper about 1-2" from the left-hand edge.
>
> Draw another line across the width of the paper about 1-2" up from the bottom. Repeat this process on the reverse side of the page.

Look at the highly effective result. You have ample room for notes, a left hand margin for special emphasis items or inserting supplementary data from the textbook, a large area at the bottom for a brief summary, and a little rectangular space for just about anything you want.

3. Get the concept then the details. Too often we focus on the details and don't gather an understanding of the concept. However, if you simply memorize only dates, places, or names, you may well miss the whole point of the subject.

A key way to understand things is to put them in your own words. If you are working from a textbook, automatically summarize each paragraph in your mind. If you are outlining text, don't simply copy the author's words.

Rephrase them in your own words. You remember your own thoughts and words much better than someone else's, and subconsciously tend to associate the important details to the core concepts.

4. Ask Why? Pull apart written material paragraph by paragraph and don't forget the captions under the illustrations.

Example: If the heading is "Stream Erosion", flip it around to read "Why do streams erode?" Then answer the questions.

If you train your mind to think in a series of questions and answers, not only will you learn more, but it also helps to lessen the test anxiety because you are used to answering questions.

5. Read for reinforcement and future needs. Even if you only have 10 minutes, put your notes or a book in your hand. Your mind is similar to a computer; you have to input data in order to have it processed. *By reading, you are creating the neural connections for future retrieval.* The more times you read something, the more you reinforce the learning of ideas.

Even if you don't fully understand something on the first pass, *your mind stores much of the material for later recall.*

6. Relax to learn so go into exile. Our bodies respond to an inner clock called biorhythms. Burning the midnight oil works well for some people, but not everyone.

If possible, set aside a particular place to study that is free of distractions. Shut off the television, cell phone, and pager and exile your friends and family during your study period.

If you really are bothered by silence, try background music. Light classical music at a low volume has been shown to aid in concentration over other types. Music that evokes pleasant emotions without lyrics is highly suggested. Try just about anything by Mozart. It relaxes you.

EARTH & PLANETARY SCIENCE

7. Use arrows not highlighters. At best, it's difficult to read a page full of yellow, pink, blue, and green streaks. Try staring at a neon sign for a while and you'll soon see that the horde of colors obscure the message.

A quick note, a brief dash of color, an underline, and an arrow pointing to a particular passage is much clearer than a horde of highlighted words.

8. Budget your study time. Although you shouldn't ignore any of the material, *allocate your available study time in the same ratio that topics may appear on the test.*

TEACHER CERTIFICATION STUDY GUIDE

Testing Tips:

1. <u>Get smart, play dumb</u>. Don't read anything into the question. Don't make an assumption that the test writer is looking for something else than what is asked. Stick to the question as written and don't read extra things into it.

2. <u>Read the question and all the choices *twice* before answering the question</u>. You may miss something by not carefully reading, and then re-reading both the question and the answers.

If you really don't have a clue as to the right answer, leave it blank on the first time through. Go on to the other questions, as they may provide a clue as to how to answer the skipped questions.

If later on, you still can't answer the skipped ones . . . **Guess.** The only penalty for guessing is that you *might* get it wrong. Only one thing is certain; if you don't put anything down, you will get it wrong!

3. <u>Turn the question into a statement</u>. Look at the way the questions are worded. The syntax of the question usually provides a clue. Does it seem more familiar as a statement rather than as a question? Does it sound strange?

By turning a question into a statement, you may be able to spot if an answer sounds right, and it may also trigger memories of material you have read.

4. <u>Look for hidden clues</u>. It's actually very difficult to compose multiple-foil (choice) questions without giving away part of the answer in the options presented.

In most multiple-choice questions you can often readily eliminate one or two of the potential answers. This leaves you with only two real possibilities and automatically your odds go to Fifty-Fifty for very little work.

5. <u>Trust your instincts</u>. For every fact that you have read, you subconsciously retain something of that knowledge. On questions that you aren't really certain about, go with your basic instincts. **Your first impression on how to answer a question is usually correct.**

6. <u>Mark your answers directly on the test booklet</u>. Don't bother trying to fill in the optical scan sheet on the first pass through the test.

Just be very careful not to miss-mark your answers when you eventually transcribe them to the scan sheet.

7. <u>Watch the clock</u>! You have a set amount of time to answer the questions. Don't get bogged down trying to answer a single question at the expense of 10 questions you can more readily answer.

EARTH & PLANETARY SCIENCE

TEACHER CERTIFICATION STUDY GUIDE

COMPETENCY 1.0 UNDERSTAND GALAXIES AND STARS.

SKILL 1.1 Identify and describe characteristics of galaxies.

Spiral Galaxy: a grouping of stars arranged in a thin disk, spiraling in a geometric pattern, that has a central pivot point (nucleus) and arms radiating outward on which stars rotate around the nucleus, somewhat suggestive of the shape of a pinwheel. Spiral galaxies usually contain a great deal of interstellar gasses and dust.

Irregular Galaxy: There is no discernable pattern in the arrangement of the stars. Like a spiral galaxy, irregular galaxies tend to have a large volume of interstellar gasses and dust.

Barred Galaxy: The shape of this type of galaxy suggests a straight center core of stars joined by two or more relatively straight arms. About 30% of all galaxies are barred.

Elliptical Galaxy: The pattern of this type of galaxy centers on an elliptical shaped central mass of stars, with other stars above or below the center, giving the entire mass an overall ovoid appearance. They contain virtually not dust or gasses and rotate very slowly if at all. Most galaxies are elliptical.

SKILL 1.2 Explain the evidence for the "big bang" model.

Big Bang Theory: a theory that proposes that all the mass and energy of the universe was originally concentrated at a single geometric point and for unknown reasons, experienced a massive explosion that scattered the matter throughout the universe.

The concept of a massive explosion is supported by the distribution of background radiation and the measurable fact that the galaxies are moving away from each other at great speed. The Universe originated around 15 billion years ago with the "Big Bang," and continued to expand for the first 10 billion years. The universe was originally unimaginably hot, but around 1 million years after the Big Bang, it cooled enough to allow for the formation of atoms from the energy and particles. Most of these atoms were hydrogen and they comprise the most abundant form of matter in the universe. Around a billion years after the Big Bang, the matter had cooled enough to begin congealing into the first of the stars and galaxies. Some of the missing pieces of the puzzle may be found in extraterrestrial objects. Prevalent cosmic theory holds that all the planets and celestial bodies formed around the same time. By absolute dating meteorite fragments found on the Earth, we have been able to gather additional information and develop a better estimate of the Earth's age. The widely accepted approximate age of the Earth is estimated as 4.6 billion years old. This age is further supported by evidence collected during the United States' exploration of the moon. Moon rocks, which range in age from 3.3 to 4.6 billion years old, give further support to the cosmic theory that our solar system- including the Earth and the Moon- was formed about the same time and by the same processes.

TEACHER CERTIFICATION STUDY GUIDE

SKILL 1.3 Know that the Sun is a typical star powered by nuclear reactions, primarily the fusion of hydrogen to form helium.

The Sun is intensely hot. At the center, it has a 140,000-kilometer diameter Core composed of hydrogen (92%) and helium (7.8%) that provide the fuel for the Sun's nuclear reaction (fusion). At approximately 15 million °C, the core gives off a tremendous amount of energy. However, the density of the Sun precludes the direct release of all this energy into space. Instead, it is slowly absorbed and re-emitted by the various layers of the Sun.

SKILL 1.4 Describe the process of the nuclear synthesis of chemical elements and how accelerators simulate the conditions for nuclear synthesis (i.e., in stars and in the early universe).

The energy of the stars originates through nuclear fusion processes. Nuclear Fusion is the process in which hydrogen atoms fuse together to form helium atoms, releasing massive amounts of energy during the fusion. It's the fusion of atoms, not combustion, which causes the star to shine. For stars like the sun, which have internal temperatures less than fifteen million Kelvin, the dominant fusion process is proton-proton fusion. For more massive stars, which can achieve higher temperatures, the carbon cycle fusion becomes the dominant mechanism. The main theme of the carbon cycle is the adding of protons, but after a carbon-12 nucleus fuses with a proton to form nitrogen-13, one of the protons decays with the emission of a positron and a neutrino to form carbon -13. More proton captures and neutron decays occur until oxygen-16 is produced and emits an energetic alpha particle to return to carbon-12 to repeat the cycle. This last reaction is the main source of energy in terms of fueling of the star. For older stars, which are collapsing at the center, the temperature can exceed one hundred million Kelvin and a third fusion process called the triple-alpha process.

Yet another class of nuclear reactions is responsible for the nuclear synthesis of elements heavier than iron. Up to iron, fusion yields energy and thus can proceed. But since the "iron group" is at the peak of the binding energy curve, fusion of elements above iron absorbs energy. Given a neutron flux in a massive star, heavier isotopes can be produced by neutron capture. Isotopes so produced are usually unstable, so there is a dynamic balance, which determines whether any net gain in mass number occurs. There is sufficient neutron capture to create isotopes up to bismuth-209, the heaviest known stable isotope. It is referred to as the "s-process" by astronomers, from "slow" neutron capture.

Current opinion is that isotopes heavier than 209Bi must be formed in the cataclysmic explosions known as supernovae. In the supernova explosion, a large flux of energetic neutrons is produced and nuclei bombarded by these neutrons build up mass one unit at a time to produce the heavy nuclei. This process proceeds very rapidly and is called the "r - process" for "rapid neutron capture". The layers containing the heavy elements may be blown off by the supernova explosion, and provide the raw material of heavy elements which condense to form new stars.

Particle accelerators use electric fields to propel electrically charged particles to speeds sufficient to cause nuclear reactions. Linear high-energy accelerators apply an alternating high-energy field to a linear array of plates. Particles approaching a plate are further accelerated towards it by the opposite charge applied to the plate. Once the particles have passed through a hole in the plate, the polarity of the particle is switched and the plate then repels the particle towards the next plate, where the same process occurs.

Black holes serve as extremely powerful and natural particle accelerators. Magnetic fields that surround black holes accelerate particles and induce high-speed collisions. These collisions produce the gamma rays characteristic of black holes. The magnetic field associated with black holes extends far beyond boundaries of the black hole, and accelerated particles that escape the black hole's gravity are further accelerated as they travel outwards. In this way, some protons are capable of reaching energies of 1000 TeV.

SKILL 1.5 Compare the use of visual, radio, and X-ray telescopes to collect data that reveal how stars differ in their life cycles.

Active Optics: a type of optical device that is composed of hexagonal pieces of mirror whose positions are controlled by a computer. Also referred to as Active Telescopes. Collectively, smaller pieces of mirror weigh less than a single large mirror and more important, they generally do not suffer from sagging problems. Small hexagonal shaped pieces of mirror are arranged next to each other to form a larger reflection surface. Computer-controlled thrusters mounted underneath the pieces control the mirror position and focus. The use of the smaller pieces working positioned so as to work in conjunction, eliminates sagging and an uneven heating and cooling problems found in extremely large, single mirror piece type telescopes.

Refractor Telescopes

Refraction: the bending of light. Example: Put a straw in a clear glass of water. Now look at straw through the side of the glass. It will appear to bend at the point where the straw enters the water. A Refractor Telescope is an optical device that makes use of lenses to magnify and display received images. Professional astronomers do not use Refractor telescopes because of two main problems: first the telescopes are affected by chromatic aberrations which make it difficult to focus on the stars, and second, because they rely solely on lenses, the telescopes have inherent weight and size restrictions.

Reflector Telescopes

Reflection: the re-emission of light of/ofan object struck by the light. Example: Look at yourself in the mirror. What your eyes see is the re-emission of light waves that have struck you *and* the mirror. A Reflector Telescope is an optical device that makes us of a mirror or mirrors, to reflect light waves to an eyepiece (an ocular), thereby eliminating chromatic aberrations. There are different types of reflector telescopes; some use mirrors only, and others make use of both lenses and mirrors.

Hubble Space Telescope: Named in honor of American astronomer Edwin Hubble, who had proved the theory of an expanding universe, the **Hubble Space Telescope** although only possessing a 2.4 meter diameter reflective surface, isn't affected by atmospheric constraints, and as a result, it provides a much clearer, higher resolution image of stellar objects than is possible through the use of an Earth-based telescope.

CCD-Charged Coupled Devices

A CCD is a camera plate made up of thousands of tiny pixels. The pixels carry a slight electrical charge and when photons strike a pixel, electrons are released. The release of the electrons causes a flow of current through an attached wire, and this current is detected by a computer chip and used to construct images based on the number of strikes. The number of strikes also shows the intensity of the received image.

Radio Telescopes

Variances in radio waves received from space can be translated into usable astronomical data. The advantages offered by use of radio telescopes are many: it's cheaper to build a radio telescope than optical telescopes, they can operate 24 hrs a day and be built just about anywhere on Earth, and they open up an entirely new window of space investigation. But they initially had one major disadvantage: the useable radio waves received from space were not overly abundant and generally very weak, and you needed a huge receiving dish to detect the signals. To overcome these problems, scientists developed a technique called Radio Interferometry. Radio Interferometry is a method of amplifying weak radio waves by lying out in a Y-shaped pattern, a series of small radio telescopes all pointed at the same point in the sky. The telescopes add their received signals together to form an overlay of signals. Computers control the angle of incidence and correlate the incoming signals. This improves resolution and limits the size of the radio telescope dish needed for a single unit. We measure the intensity of the light by the ratios of the apparent presence or absence of colors. **Intensity:** the amount of light contained in a space. Intensity varies by distance. The further away you are, you will see a drop in intensity equal to 1 over the distance squared. Intensity gives off a continuous amount of colors, but the intensity of the colors seen varies in accordance with the temperature of the object. Stars follow the same principles of emission. We observe the intensity of the stars by using a red & blue filter on a photon counter mounted on a telescope. The red and blue ratio determines the color index. From the color index, we can determine the temperature, size, properties, and material composition of the star.

X-rays are a form of electromagnetic radiation with a wavelength in the range of 10 to 0.01 nanometers. X-ray detectors measure x-ray photons that react with the detector, producing a charge. Over time, these devices accumulate enough measurements of individual photons to produce an accurate measurement of the total source. X-ray photons have a higher energy than optical photons, so these individual photons are easier to detect. Currently, most x-ray telescopes use CCD x-ray detectors.

X-ray emission is the principal means of energy loss from most stellar coronas. By recording changes in a star's x-ray emission over extended amounts of time, x-ray telescopes can help scientists to study stellar life cycles because stars at different stages of the life cycle are known to emit x-rays in different amounts. Steady x-rays emitted from stars are thermal, meaning the x-ray spectrum is characterized by temperature. In older pulsars, where the neutron star has had time to cool and the amount of energy released by the current that flows to the star's surface is much less than that of a younger star, x-rays are entirely pulsed, with no thermal component.

X-rays also enable scientists to study black holes, which are formed by the gravitational collapse of massive stars. When stars approach black holes, gaseous material of the star's surface is sucked into the black hole and elevated to temperatures of millions of degrees by extreme speed and friction. These temperatures produce x-rays, which can be detected on Earth. X-rays emitted from black holes can indicate the final stage in a star's life cycle.

SKILL 1.6 Describe, in terms of color and brightness, how the evolution of a star is determined by a balance between gravitational collapse and nuclear fusion.

Stars form in Planetary Nebulae: cold clouds of dust and gas within a galaxy, and go through different stages of development in a specific sequence. This theory of star development is called the Condensation Theory.

Sequence of Development
In the initial stage, the diffuse area of the nebula begins to shrink under the influence of its own weak gravity. The cloudlike spheres condense into a knot of gasses called a **Protostar.** The original diameter of the protostar is many times greater than the diameter of our solar system, but gravitational forces causes it to continue to contract. This compression raises the internal temperature of the protostar. When the protostar reaches a temperature of around 10 million degrees C, nuclear fusion starts, which stops the contraction of the protostar and changes its status to a star.

Nuclear Fusion: the process in which hydrogen atoms fuse together to form helium atoms, releasing massive amounts of energy during the fusion.

It's the fusion of atoms, not combustion, which causes the star to shine. A star's life cycle depends on its initial mass. Red stars have a small mass. Yellow stars have a medium mass. Blue stars have a large mass. Large mass stars consume their hydrogen at a faster rate and have a short life cycle in comparison to small mass stars that consume their hydrogen at a much slower rate. All stars eventually convert a large percentage of their hydrogen to heavier atoms and begin to die. However, just as their mass determines the length of their life, it also determines the pattern they follow in the last stages of their existence.

Lower Main Sequence Stars

When small and medium mass stars (such as the Sun) consume all of their hydrogen, their inner cores begin to cool. The stars begin to consume the heavier elements produced fusion (carbon and oxygen) and the star's shell tremendously expands outward, causing the star to become a **Giant Star**: large, cool extremely luminous stars 10 to 100 times the diameter of the Sun. Example: In roughly 4.6 billion years, our Sun will become a Giant. As it expands, its outer layers will reach halfway to Venus. The dying Giant gives off thermal pulses approximately every 200,000 years, throwing off concentric shells of light gasses enriched with heavy elements. As it enters its last phases of the life cycle its depleted inner core begins to contract, and the Giant becomes a **White Dwarf Star**: a small, slowly cooling, extremely dense star, no larger than 10,000 km in diameter. The final phase of a lower main sequence star life cycle can take two paths: most main sequence white dwarfs after a few billion years completely burn out to become **Black Dwarfs**: cold, dead stars. However, if a White Dwarf is part of a **Binary Star**: two suns in the same solar system, instead of slowly cooling to become a Black Dwarf, it may capture hydrogen from its companion star. If this happens, the temperature of the White Dwarf soars and when it reaches approximately 10 million degrees C, a nuclear explosion occurs, creating a Nova: a sudden brightening of a lower main sequence star to approximately 10,000 times its normal luminosity; caused by the explosion of the star. A nova reaches its maximum brightness in a short time (one or two days) and then gradually dims as the gasses and cosmic dust cool.

Upper Main Sequence Stars

The initial sequence of the high mass, upper main sequence stars is identical to the lower mass stars, Planetary Nebulae to Protostar. However, if the protostar accretes enough material, it forms as a Blue Star. When a Blue Star has consumed all of its hydrogen it too expands outward, but on a much larger scale then experienced by a lower mass star. It becomes a **Supergiant Star**: an exceptionally bright star, 10 to 1000 times the diameter of the Sun. The Supergiant's now depleted core cannot support such a vast weight and collapses inward causing its temperature to soar. When it reaches roughly 599 million degrees C, it implodes and then explodes, creating a Supernova: the massive explosion of an upper main sequence Supergiant star; caused by the detonation of carbon within the star. A supernova releases more energy than Earth's sun will produce in its entire life cycle. The luminosity of a supernova is as bright as 500 million Suns.

Example: Chinese astronomers in 1054 recorded the sudden appearance of a new star in what is now know as the Taurus Constellation. Bright enough to be see during daytime for over a month, it remained visible for over 2 years. 90 percent of the shattered mass scatters into space, becoming planetary nebulae from which the life cycle may begin anew. The other 10 percent, the core of the star, is blown inward, becoming a Neutron Star: a very small - 10 km diameter core of a collapsed Supergiant star that rotates at a high speed (60,000 rpm) and has a strong magnetic field. A neutron star may capture gas from space, a companion star, or a nearby star and become a Pulsar: a neutron star that emits a sweeping beam of ionized-gas radiation. As the pulsar rotates, the beams of light sweep into space similar to a beacon from a lighthouse. Since first discovered in 1967, over 350 pulsars have been catalogued. The alternate product of a supernova is a **Black Hole**: a volume of space from which all forms of radiation cannot escape. Black Holes are created when a Supergiant star with a mass roughly 3 times that of the Sun implodes.

COMPETENCY 2.0 UNDERSTAND SOLAR SYSTEMS.

SKILL 2.1 Explain how the solar system was formed, including differences and similarities among the sun, terrestrial planets, and the gas planets, and cite the evidence from Earth and moon rocks that indicate that the solar system was formed approximately 4.6 billion years ago.

Formation of Earth and the Solar System

Most cosmologists believe that the Earth is the indirect result of a supernova. The thin cloud (planetary nebula) of gas and dust from which the Sun and its planets are formed, was struck by the shock wave and remnant matter from an exploded star(s) outside of our galaxy. In fact, the stars manufactured every chemical element heavier than hydrogen. The turbulence caused by the shock wave caused our solar system to begin forming as is absorbed some of the heavy atoms flung outward in the supernova. In fact, our solar system is composed mostly of matter assembled from a star or stars that disappeared billions of years ago. Nebulae spun faster as it condensed and material near the center contracted inward. As more materials came together, mass and consequently gravitational attraction increased, pulling in more mass. This cycle continued until the mass reach the point that nuclear fusion occurred and the Sun was born. Concurrently, the Proto-sun's gravitational mass pulled heavier, denser elements inward from the clouds of cosmic material surrounding it. These elements eventually coalesced through the process of Accretion: the clumping together of small particles into large masses, into the planets of our solar system. The period of accretion lasted approximately 50 to 70 million years, ceasing when the protosun experienced nuclear fusion to become the Sun. The violence associated with this nuclear reaction swept through the inner planets, clearing the system of particles, ending the period of rapid accretion. The closest planets Mercury, Venus, and Mars, received too much heat and consequently, did not develop the planetary characteristics to support life, as we know it. The farthest planets did not receive enough heat to sufficiently coalesce the gasses into solid form. Earth was the only planet in the perfect position to develop the conditions necessary to maintain life.

Prevalent cosmic theory holds that all the planets and celestial bodies formed around the same time. By absolute dating meteorite fragments found on the Earth, we have been able to gather additional information and develop a better estimate of the Earth's age. The widely accepted approximate age of the Earth is estimated as 4.6 billion years old. This age is further supported by evidence collected during the United States' exploration of the moon. Moon rocks, which range in age from 3.3 to 4.6 billion years old, give further support to the cosmic theory that our solar system- including the Earth and the Moon- was formed about the same time and by the same processes.

TEACHER CERTIFICATION STUDY GUIDE

SKILL 2.2   Know the current evidence for the existence of planets orbiting other stars.

Extrasolar planets are detected through the influence they enact on their host stars. A planet orbiting a star causes the star to wobble, changing the wavelength of the star's light that is detected on Earth. This change is known as the Doppler effect. Using this method of detection, during the past 15 years over one hundred planets have been determined to orbit stars outside of our solar system.

The first extrasolar planets were detected in 1991, when scientists discovered three objects orbiting the star PSR B1257+12. One of these planets is Moon-sized, while the other two are 2 to 3 times the size of Earth. PSR B1257+12 is a dying pulsar star left over from a supernova explosion in the constellation Virgo, and emits radiation that would prevent any form of life in its planetary system. Its three planets were detected by measuring variations on the pulsation speed of the pulsar. These variations are assumed to be gravitational effects of the three planets.

In 1995, a slight perturbation in the position of the star 51 Plegasi was interpreted as evidence of the presence of another extrasolar planet. 51 Plegasi is located in the constellation of Pegasus, and is very similar to our Sun in temperature, size, rotation speed and emitted radiation. This star is a spectral type G2-3 V main-sequence located 42 light-years from Earth. Using a high-resolution spectrograph, scientists have recently discovered that the star's line-of-sight velocity changes approximately 70 meters per second every 4.2 days (according to a study by Michel Mayor and Didier Queloz of Geneva Observatory), indicating that the orbiting planet lies a mere 7 million kilometers from its host star. This distance, much smaller than that of Mercury to the Sun, is too close to be life-conducive. The planet of 51 Plegasi is at least half the mass of Jupiter.

It is now believed that the star Gl229 has an object 20 times the size of Jupiter orbiting at a distance of 44 AU. This object is most likely a brown-dwarf rather than an ordinary planet. Evidence also indicates that the pulsar PSR 0329+54 (PKS B0329+54) is orbited by a planet greater than 2 times Earth's mass with a period of 16.9 years. It has also been shown that the star HR3522 has an orbiting planet approximately 0.8 times the size of Jupiter.

The star Beta Pictoris may also have several orbiting planets. Beta Pictoris is surrounded by a disk of gas and dusk (protoplanetary disk) that is considerably thinner than previously thought, indicating that planets may have formed through gas accretion in the disk. This disk is also known to be warped from the gravitational influence of orbiting planets.

EARTH & PLANETARY SCIENCE

In 1996, planets were discovered to orbit the stars 70 Virginis and 47 Ursae Majoris. 70 Virginis is a G5V (main sequence) star 78 light-years from Earth. This star is very similar to the Sun, but may be more than 3 billion years older. Its planet orbits with a period of 116 days and has a mass approximately 9 times that of Jupiter. The temperature of this planet is believed to be approximately 85 degrees Celsius, and may permit the existence of water and complex organic molecules. 47 Ursae Majoris is a G0V star about 44 light-years away. Its planet has a period of over three years, and a mass approximately three times that of Jupiter.

SKILL 2.3   Describe changes in the solar system over time.

According to the nebular theory, the Solar System was formed by a gravitational collapse of a giant molecular cloud approximately 4.6 billion years ago. From the time of its formation, the Solar System has undergone constant change, often as a result of planetary impacts and collisions.

Inner Solar System Change:

The Sun, Mercury, Venus, Earth and Mars currently make up the inner solar system. However, upon the formation of the solar system, it is believed that five planets existed in this inner orbit of the Sun. According to this theory, known as the "Giant Impact Hypothesis," the current inner solar system was established when Earth collided with a fifth planet, or Mars-sized body, and the consequential impact of this collision produced the Moon. Following this impact, the fifth planet is said to have drifted out of the orbit of the inner solar system, leaving the current four planets.

With respect to recent, more specific changes of the inner solar system, the Sun's magnetic field has become 230 % stronger during the past century, and its overall energetic activity continues to increase. The icecaps once found on Mars have melted, causing drastic changes in its surface features and increasing the atmospheric density by 200 % since 1997.

The surface of Earth grew dimmer by 4-6 % around 1960, and then began to brighten once again around 1994. This recent increase has been attributed to global warming and the greenhouse effect. The Earth's icecaps have decreased by 40 % during the past 40 years, and in 1997, the structure of the Earth began to change from its original, egg-like shape of elongation at the poles, to a more "pumpkin" shape with flattening at the poles.

Asteroid Belt Change:

The asteroid belt is the region of the solar system between the planets Mars and Jupiter where the greatest concentration of asteroid orbits can be found. According to the nebular theory, upon formation, the asteroid belt contained enough matter to make up a planet. The planetesimals within the asteroid belt were never able to form a protoplanet, however, because of the rapid increase in size of the protoplanet Jupiter. Due to the orbital resonances created by the massive size of Jupiter, the matter found within the asteroid belt was unable to condense, and was either scattered or held in narrow orbital bands. Eventually these orbital resonances scattered over nine tenths of the original matter found in the asteroid belt.

Change in the Outer Planets:

After the initial formation of Solar System, outer protoplanets continued to increase in size and began to acquire gas from the protoplanetary disk, the rotating body of dense gas surrounding a newly formed star that may eventually evolve into a planetary system. The relative rates at which the protoplanets reached the critical mass required to capture gas from the protoplanetary disk are attributed to the positions of the planets in the disk. Thus Jupiter, the most interior protoplanet, was the first newly formed planet to begin to capture helium and hydrogen gases, as its closer position to the Sun created a higher orbital speed, disk density and collision frequency. Saturn began to accrete gas from the protoplanetary disk shortly after Jupiter. Uranus and Neptune reached the critical mass to acquire gas significantly later, explaining their smaller sizes.

The Kuiper belt is the region of the Solar System between the Sun and Neptune. As the gas capture and gravity of the outer planets increased, objects within the Kuiper Belt were scattered inwards by Saturn, Uranus and Neptune, and outwards by Jupiter, resulting in the migration of Jupiter toward the Sun, while Saturn, Uranus and Neptune migrated away from the Sun.

More recent changes to the outer planets of the Solar System are numerous. Since 1992, the size of Jupiter's magnetic field has more than doubled, and the planet itself has become so highly energized that a glowing ring of energy now surrounds it. Saturn's magnetic field has been increasing as well, and its polar regions have been brightening. Saturn's weather changed drastically between 1980 and 1996, when the speed of rotation of its clouds was reduced by 58.2 %. Both Uranus and Neptune have experienced recent magnetic pole shifts, and since 1996, Neptune has become 40 % brighter in infrared. Additionally, Neptune's moon demonstrates a recent and large increase in atmospheric pressure and temperature. Finally, as of 2002, Pluto had experienced a 300 % increase in atmospheric pressure since 1988, and has become noticeably darker.

COMPETENCY 3.0 UNDERSTAND PLANETS AND SATELLITES.

SKILL 3.1 Cite various forms of evidence that indicate the proximity of the planets in the solar system in relation to Earth and the stars.

Historical Evidence:

Johannes Kepler, born in 1571, was the first person to propose a method for determining the relative distances of the planets from the Sun. Kepler claimed that the orbits of the eight planets around the Sun were elliptical, and that the closer a planet was to the Sun, the faster the planet traveled and therefore completed its orbit around the Sun. Using this information, Kepler was able to determine how much closer or farther other planets were to the Sun than was Earth. Kepler observed that Mars required less than two years to orbit the Sun, while Saturn required 29 years to complete its orbit. Comparing these periods of time to that of Earth's rotation around the Sun, Kepler correctly concluded that Mars was approximately 1.5 times farther from the Sun than was Earth, where Saturn was 10 times farther away.

In 1672, the astronomer Gian Domenica Cassini measured the distance to Mars using a method called parallax. Parallax is the apparent shift of an object against a background, such as a star or planet, caused by a change in observer position. Astronomers during this time were able to roughly calculate the distance to certain planets from Earth by recording the distance between two points from which an object was observed, as well as the parallax of this object.

Contemporary Methods:

Today, scientists use parallax to calculate the exact distances to the planets of our solar system, as well as the distances to distant objects, such as stars. Star distances are determined by observing stars from two locations along the Earth's orbit around the Sun, instead of from two locations on the Earth. By taking these measurements several months apart, scientists are able to have a separation of observer positions over 100 million miles apart.

Advanced technology also allows scientists to measure distances to other planets more directly. Spacecrafts sent to other planets are capable of sending radio signals back to Earth. By knowing the speed at which the signal travels and the time it takes for the signal to reach Earth, scientists are able to calculate the distance between the other planet and Earth. Today it is also possible to send powerful radar signals of known speed to another planet, and then record how long it takes for the signals' echo to return to Earth.

Distances of Planets:

By assuming the orbits of each planet are circular, the planet to Sun distances can be determined using the before-mentioned methods.

Planet	Distance from Sun in AU
Mercury	0.39
Venus	0.72
Earth	1.0
Mars	1.5
Jupiter	5.2
Saturn	9.5
Uranus	19.2
Neptune	30.1
Pluto	39.5

AU = distance from Earth to the Sun = 150 million kilometers or 93 million miles

Distances between the orbits of Earth and another planet, such as Earth and Venus, for example, can be determined by subtracting the distance from Venus to the Sun (0.72 AU), from the distance of Earth to the Sun (1 AU).

Distances of Stars:

1 light year = about 63,240 astronomical units

Common Name	Scientific Name	Distance (light yrs)
Sun		
Proxima Centauri	V645 Cen	4.2
Rigil Kentaurus	Alpha Cen A	4.3
	Alpha Cen B	4.3
Barnard's Star		6.0
Wolf 359	CN Leo	7.7
	BD +36 2147	8.2
Luyten 726-8A	UV Cet A	8.4
Luyten 726-8B	UV Cet B	8.4
Sirius A	Alpha CMa A	8.6
Sirius B	Alpha CMa B	8.6
Ross 154		9.4
Ross 248		10.4
	Epsilon Eri	10.8
Ross 128		10.9
	61 Cyg A (V1803 Cyg)	11.1
	61 Cyg B	11.1
	Epsilon Ind	11.2
	BD +43 44 A	11.2
	BD +43 44 B	11.2
Luyten 789-6		11.2
Procyon A	Alpha CMi A	11.4
Procyon B	Alpha CMi B	11.4
	BD +59 1915 A	11.6
	BD +59 1915 B	11.6
	CoD -36 15693	11.7

SKILL 3.2   Cite various forms of evidence that Earth and other planets change over time.

Earth: Climate

Since its formation some 4.6 billion years ago, Earth has experienced drastic climate changes, such as the periods of reduced temperatures known as the Ice Ages. Ice Ages generally result in mass glaciation, which is the expansion of polar and continental ice sheets, as well as mountain glaciers. In reference to the last million years, an Ice Age is a period of colder temperatures during which extensive ice sheets are found over the North American and Eurasian continents.

Thus far, the Earth has experienced four major ice ages. The earliest ice age is believed to have occurred around 2.7 to 2.3 billion years ago during the early Proterozoic Eon. Another ice age occurred during the Crogenian period from 800 to 600 million years ago (mya). A minor ice age occurred from 460 to 430 mya during the Late Ordovician Period. And finally, the Karoo Ice Age lasted from 350 to 260 mya, when extensive polar ice caps appeared at intervals during the Carboniferious and early Permian Periods.

Earth is currently experiencing an Ice Age that began 40 mya with growth of an ice sheet in Antarctica. Around 3 mya, this Ice Age intensified as sheets began to spread in the Northern Hemisphere. Since this time, Earth has experienced cycles of glaciation on 40,000 to 100,000-year time scales. In between ice ages and during times of warmer temperatures within ice ages, extended periods of temperate and/or tropical temperatures may exist. These times are known as interglacial periods. We are currently experiencing an interglacial period during which the Earth's icecaps have decreased by 40 % over the past 40 years.

Earth: Shape

In the past century, the shape of Earth has begun to change from its "egg-like" form to a more "pumpkin-like" structure. Once more elongated at the poles, Earth has begun to flatten at its North and South extremities and widen at the equator, a change scientists have recently attributed to climate events, such as the El Nino weather pattern, which effect the balance of water found in the oceans, continents and atmosphere.

NASA scientists Cheng and Tapley discovered that two large increases in the bulging of the Earth at the equator occurred during El Nino-Southern Oscillation events from 1986-1991 and 1996-2002. During these periods, heavy rains caused by warmer waters moved into the central Pacific Ocean. Because of this shift in the water balance of the Earth, Peru experienced flooding, while Australia experienced widespread drought. Scientists believe that the change of the Earth's shape during this time is associated with the larger water mass found near the equator, showing the connection between Earth's transformation and large-scale weather events.

Other Planets

Other planets within our solar system also demonstrate continual change. For example, the Sun's magnetic field has become 230 % stronger during the past century, and its overall energetic activity continues to increase. The icecaps once found on Mars have melted, causing drastic changes in its surface features and increasing the atmospheric density by 200 % since 1997. Since 1992, the size of Jupiter's magnetic field has more than doubled, and the planet itself has become so highly energized that a glowing ring of energy now surrounds it. Saturn's magnetic field has been increasing as well, and its polar regions have been brightening. Saturn's weather changed drastically between 1980 and 1996, when the speed of rotation of its clouds was reduced by 58.2 %. Both Uranus and Neptune have experienced recent magnetic pole shifts, and since 1996, Neptune has become 40 % brighter in infrared. Additionally, Neptune's moon demonstrates a recent and large increase in atmospheric pressure and temperature. Finally, as of 2002, Pluto had experience a 300 % increase in atmospheric pressure since 1988, and has become noticeably darker.

SKILL 3.3 Describe the influence of collisional processes on early Earth and other planetary bodies in terms of shaping planetary surfaces and affecting life on Earth.

Our planet's earliest history was quite violent. Besides being repeatedly struck by meteorites, the Earth's birthing process produced frequent eruptions that caused water- primarily in the form of water vapor from gaseous hydrous minerals- to rise to the surface. The solar system contributed additional water as the Earth acquired its basic form. Comets- primarily composed of water and other light, hydrous materials- bombarded the Earth's surface, and their melting added to the water volume. The primordial seas were initially very acidic, but as the hydrologic cycle of evaporation and precipitation took hold, soluble minerals dissolved and were carried to the seas, gradually easing the degree of acidity. The exchange of minerals though the hydrologic cycle further modified the water's chemistry. Sodium, being extremely soluble, remained in the water longer than the other common elements, and the percentage of sodium in the Earth's oceans has not varied appreciably foe at least 600 million years.

COMPETENCY 4.0 UNDERSTAND TECTONIC PROCESSES

SKILL 4.1 Diagram the major divisions of the geologic time scale as a basis for understanding changes in the Earth's processes.

In the Earth Sciences, when you talk about time, you must think in terms of huge expanses of time.

Geological Time Scale: the calendar/clock of events in geology based on the appearance and disappearance of fossil assemblages.

The scale is divided into time units that are given distinctive names and approximate start and stop dates. These dates are based upon a reexamination of previously discovered fossils, using absolute dating techniques.

Eons: The largest scale division
Eras: Divided into sub categories based on profound differences in fossil life.
Periods: Smaller divisions within eras based on less profound differences in the fossil record.
Epochs: Sub-categories within some periods that are specific to the types of fossils found within.
Pre-Cambrian Time: Comprised of the Hadean, Archean, and Proterozoic Eons, 87% of all geologic time is considered Pre-Cambrian.

Major Geological Time Scale Divisions

Eon, Era, Period, or Epoch	Time Start (mya)
Hadean Eon	4600
Archean Eon	3800
Early Era	3800
Middle Era	3400
Late Era	3000
Proerozoic Eon	2500
Paleo-proterozoic Era	2500
Meso-proterozoic Era	1600
Neo-proterozoic Era	1000
Phanerozoic Eon	570
Paleozoic Era	570
Cambrian Period	570
Ordovician Period	505
Silurian Period	438
Devonian Period	408
Carboniferous Period, includes the:	360
Mississippian Period	360
Pennsylvanian Period	320
Permian Period	286
Mesozoic Era	245
Triassic Period	245
Jurassic Period	208
Cretaceous Period	144
Cenozoic Era	66
Tertiary Period, comprised of:	66
Paleogene Period	66
Paleocene Epoch	66
Eocene Epoch	58
Oligocene Epoch	37
Neogene Period	24
Miocene Epoch	24
Pliocene Epoch	5
Quartenary Period	2
Pleistocene Epoch	2
Holocene Epoch	10,000 years

SKILL 4.2 Describe how earthquake intensity, magnitude, epicenter, focal mechanism, and distance are determined from a seismogram.

Seismic waves are measured using a device called a Seismometer, a type of motion sensor. The seismometer is anchored to the earth and a heavy weight is suspended on its frame. As the Earth materials move, the weight also moves and electronically sends a signal to a recording device called a Seismograph. Movements are displayed as a series of lines on a recording chart called a Seismogram, reflecting the seismic energy detected at a particular location.

Types of Seismic Waves:

P-Wave (Primary Wave): Also sometimes called a "Push-Pull Wave or Compression Wave," the P-Wave moves through both solids and liquids. The P-Wave has a pulsating, "push-pull" type motion. It compresses material as it moves through it. The fastest moving of the seismic waves (4-7 Km/sec), the P-Wave is the first wave to reach the seismometers.

S-Wave (Secondary Wave) (Shake Wave): Moves only through solid material. Always shorter than a P-Wave, the S-Wave (2-5 Km/sec) is the second wave that reaches the seismometer. Motion is a sinuous "side to side" movement.

L-Wave (Surface Wave): The L-Wave is much slower than either the P or S-Waves, but creates lots of ground movement. Because it is slower, the L-Wave takes longer to pass a location and consequently, the intense, undulating ground motion creates the greatest amount of damage in an earthquake. The L-Wave undulates with a rolling motion of the earth, similar to ocean waves.

Measuring Magnitude and Intensity:

Magnitude: the relative measure of how big an earthquake is; how much energy is released.

Intensity: the measure of observable effect in terms of damage and destruction, caused by an earthquake.

Richter Scale: the primary scale used by seismologists to measure the magnitude of the energy released in an earthquake. It is a logarithmic scale.
    Example: A 5.0 earthquake is 10 times stronger than a 4.0 earthquake, and 100 times stronger than a 3.0 earthquake.

Although the Richter Scale is still the most widely known scale, it is only one of several types of scales used to measure magnitude.
    Example: The Fuji scale is gaining popularity a being a more accurate measurement of the energy released by an earthquake.

A series of seismometers are used to locate the epicenter of an earthquake through a geometrical process called triangulation. A minimum of three seismometers is required to accurately triangulate the epicenter. Seismometers are also used to determine the magnitude and distance by plotting a travel-time curve, derived by measuring the time lag between to arrival of the P and S waves.

SKILL 4.3 Compare major types of volcanoes in terms of shape and chemical and rock composition.

Shield Volcano: This type of volcano forms when the magma is silica poor. Although the magma is runny, it is still thick enough to pile up in layers and form the volcanic cone. The succeeding layers eventually build upwards to form a massive volcanic mountain.

The cone is tall (up to 9,000 meters), broad across, and composed of overlapping layers of lava flow. The lava cools quickly and resembles asphalt in appearance. Because shield volcanoes are produced over hot spots and at spreading centers, the broadest and tallest parts are usually underwater. Example: Mauna Loa in Hawaii.

The eruptions are characteristically very quiet with lots of lava and little ash. The lava travels slowly unless moving downhill. These eruptions are generally non-lethal because there is plenty of time to get out of the way of the slow moving lava. The eruptions occur over a long period of time (years). Two types of lava erupt from a shield cone, AA and Pahoehoe:
- AA has sharp edges.
- Pahoehoe has smooth edges and looks like strands of black rope glued together. Gasses escape from fissures all around the volcano.

Cinder Cone Volcano: This type of volcano forms from thick, silica rich magma that does not easily flow. However, it is the local concentrations of gas in the magma that cause the eruptions that build cinder cones, not the silica content. Gas builds up pressure in the neck of the volcanic tube. The gas pressure eventually builds to the point that it overcomes the resistance offered by the surrounding rock materials and suddenly blows off the top of the cone. A huge mass of liquid magma and Pyroclastic Rock: a rock formed from explosively erupted rock particles or magma, is flung outward in a violent explosion. Accompanying the pyroclastic ejecta is a huge release of gas that forms into a glowing gas cloud that moves rapidly down the side of the volcano. This gas cloud, the Nuée Ardente: a glowing, highly heated mass of gas-charged lava, travels at 600 mph when first expelled and is still moving at 200 mph when it reaches the bottom of the cone. The Nuée Ardente is very deadly. Besides being scorching hot, it causes a smothering effect by displacing or burning up all oxygen in its path. The Nuée Ardente also creates extremely strong winds that can affect areas as much as 20 miles away from the volcano. You also get

Lahars: a flowing slurry of volcanic debris and water. These mudflows have the consistency of wet concrete and can cause widespread devastation.

A cinder cone volcano is short (less than 500 meters tall), very small in width, and is composed of layers of pyroclastic debris.
Example: Paricutín, in Mexico.

Composite Cone Volcano (Stratovolcano): This type of volcano is part Shield Volcano and part Cinder Cone. The cone is composed of layers of basalt intermingled with layers of pyroclastic debris. The layers are generally laid down like a layer cake, a layer of shield eruption, and then a layer of cinder eruption. A large number of volcanoes are classified as composite volcanoes, and they form some of the most beautiful and spectacular volcanic vistas. Stratovolcanoes vary in size from 100 to 3500 meters, and grow from the many eruptions that take place over long periods of time. They tend to have steep, but unconsolidated slopes, reflecting the composite composition of the cone materials.
Example: Mt. St. Helens or Mt. Rainier in Washington State, or Mt. Vesuvius near Naples, Italy.

SKILL 4.4   Describe the location and characteristics of volcanoes that are due to hot spots and those due to subduction.

Subduction Zone: A long, narrow belt where a lithospheric plate dives into the asthenosphere.

At the location of ocean/continental boundaries, the colliding plates produce effects relatively similar to ocean/ocean collisions, but the difference in density between the materials involved causes the oceanic plate to subduct under the continental plate. Subduction forces the continental materials upward, creating a line of on-shore volcanic mountains along the subduction zone.

Hot Spot: an offshoot of a convection cell that burns through the lithospheric material in the middle, not at the edges, of a plate.

Aseismic Ridge: a series of volcanoes (underwater or above the surface) formed by the movement of the plates over a hot spot. They usually form a dogleg pattern, showing the change of direction in the plates' movement.
Example: The Hawaiian Islands.

Magma is produced at Hot Spots, Spreading Centers, and Subduction Zones and varies in composition according to where it is produced.
- Hot Spot and Spreading Center: The magma is produced in the asthenosphere and comes up mantle plume. This magma is pure magma, dense and silica poor. However overall, the material is very runny and fluid.

- Subduction Zone: The magma is impure because ocean sediment is sucked down and mixed with the magma. The silica rich, water saturated magma is thicker because of the addition of ocean sediment.

SKILL 4.5 Relate geologic structures to tectonic settings and forces.

Plate tectonic movement results from the motion induced in the lithosphere by the rise and fall of convection cell material in the asthenosphere.

Plate Boundaries: the points at which the edges of the tectonic plates abut.

Three motions characterize interactions at the plate boundaries: separation, collision, or lateral movement. Those motions directly correlate with the categorizations of plate boundaries; Divergent, Convergent, and Transform.

Divergent Boundaries:
Divergent Boundary: The plates are separating and moving away from each other.

Ocean/Ocean Boundaries: The materials involved are composed of heavy and dense, but very thin, dark colored oceanic lithospheric material, usually Basalt. As the magma rises, the ocean floor begins to dome upward. The upward pressure eventually forces an underwater rip in the center of the dome and the magma erupts. The erupted materials cool rapidly and build upward, forming Mid-Ocean Ridges, which are fairly common and found all over the globe. The ocean floor is constantly being pushed apart at these boundaries, causing of Sea Floor Spreading. This results in the creation of huge oceanic plates.

Continental/Continental Boundaries: The materials involved are composed of less dense, but very thick, lighter colored continental lithospheric material, usually Granite.

Convergent Boundaries:
Convergent Boundary: The plates are moving toward, and collide with each other.

Ocean/Ocean Boundaries: These plates are forced together by the spreading of the ocean floor. Tremendous frictional forces are created as the plates collide and some of the oceanic material builds upward, while other oceanic material bends downward.

Subduction Zone: A long, narrow belt where a lithospheric plate dives into the asthenosphere. The rate of subduction is relatively equal to the rate of formation of new oceanic lithospheric material at divergent boundary spreading centers. In effect, the ocean floor recycles itself.

Ocean/Continental Boundaries: The colliding plates produce effects relatively similar to ocean/ocean collisions, but the difference in density between the materials involved causes the oceanic plate to subduct under the continental plate. Subduction forces the continental materials upward, creating a line of on-shore volcanic mountains along the subduction zone.

Continental/Continental Boundaries: Both edges are too light to subduct. Instead, one will over ride the other, causing an uplift of material.

Transform Boundaries:
Transform Boundary: The plates move laterally to each other. As the plates grind sideways, intense frictional forces are created as the lithospheric materials try to oppose the movement. A transform boundary may be found in any location where plates abut. They may be composed of any type of lithospheric material (oceanic or continental), and they produce extreme seismic effects when the pressure between moving boundaries is released. This sudden release of pressure creates widespread destruction along the fault lines.

SKILL 4.6 Describe the evidence for plate tectonics on the sea floor and on land.

Support of Tectonic Theory:

Shape of the Continents: When graphically displayed, the continents look like they should largely fit together in a "Jigsaw" puzzle fashion.

Paleomagnetism: As igneous rock cools; iron minerals within the rock will align much like a compass, to the magnetic pole. Scientific research has shown that the magnetic pole periodically—hundreds of thousands of years—reverses polarity.
- Normal Polarity: Magnetic North.
- Reverse Polarity: Magnetic South.

Research also shows that the bands of rocks on either side of a spreading center are mirror images of each other with regards to magnetic polarity, and that the alignment of minerals indicate a periodic shift in polarity. The reversals in polarity can be visualized as alternating "stripes" of magnetic oceanic materials.

Age of the Rock: Besides being mirror images magnetically, dating research done on rocks on either side of a spreading center also indicate a mirroring of age. The age of the rock on either side of a spreading center are mirror images and get progressively older as you move away from the center. The youngest rock is always found directly at the spreading center. In comparison to continental rock materials, the youngest rock is found on the ocean floor, consistent with the tectonic theory of cyclic spreading and subduction. Overall, oceanic material is roughly 200 million years old, while most continental material is significantly older, with age measured in billions of years.

Climatology: This is one of the most compelling arguments supporting plate movement. Cold Areas show evidence of once having been hot and visa-versa. In example, Coal needs a hot and humid climate to form. It does not form in the areas of extreme cold. Although Antarctica is extremely cold, it has huge coal deposits. This indicates that at one time in the past, Antarctica physically must have been much closer to the equator.

Evidence of Identical Rock Units: Rock units can be traced across ocean basins. Many rocks are distinctive in feature, composition, etc. Identical rocks units have been found on multiple continents, usually along the edges of where the plates once apparently joined.
    Example: A significant number of South American and African rock unit formations are identical.

Topographic Evidence: Topographic features can be traced across ocean basins. Some glacial deposits, stream channels, and mountain ranges terminate on one continent near the waters edge, and resume on another continent relatively the same position.

Fossil Evidence: Limited range fossils that could not swim or fly are found on either side of an ocean basin.
    Example: The fossilized remains of pandas, kangaroos and the long extinct Metesaurus are unique to only two areas of the world. Ocean basins separate these areas.

Sea Turtle Migration: The genetic instincts of Sea Turtles drive them to return to the islands where they hatched in order to lay eggs. The migration of the sea turtles over thousands of miles is well documented. The diverse location and number of islands to which the sea turtles migrate suggests that the plate movement has changed the location of the islands from their original position; immediately off shore of major continental masses.

COMPETENCY 5.0 UNDERSTAND OCEANS.

SKILL 5.1 Describe the chemical and physical properties of sea water.

Significance of Water

Water (H_2O) is significantly different from its immediate Hydrogen compound cousins. **Compounds** are substances that contain two or more elements in a fixed proportion. Generally, the heavier molecules have higher boiling and freezing states based upon molecular weight.

A group of atoms held together by chemical bonds is called a **molecule**. The bonds form when the small, negatively charged electrons found near the outside of an atom are shared or transferred between the atoms. The bonds formed by the shared pair of electrons are known as **covalent bonds**. Most substances tend to adopt a solid or gaseous form. Water is different. It wants to be a liquid. A water molecule forms when covalent bonds are established between two hydrogen atoms and one oxygen atom. However, unlike the other hydrogen compounds, water has its two hydrogen molecules on one side of the atom. It's a polar molecule. This arrangement of atoms is based on the distribution of the water molecule's oxygen electrons. The electrons cause the geometric shape of the molecule to be angular. This angular shape makes the molecule electrically asymmetrical (polar). This polar arrangement gives water some very special properties:

- The polar molecule acts similar to a magnet. Its positive ends (the hydrogen) attracts particles having a negative charge, and its negative end (the oxygen) attracts particles that have a positive charge. This arrangement is the basis for the water molecule's rightful description as the "**Universal Solvent**." When water comes into contact with compounds- for example salts- held together by the attraction of opposite charges, the water molecule separates that compound's elements from each other.
- Another unique property of water is that water likes itself; it has a natural tendency to stick to itself. Once again, this property is based upon the polar nature of the water molecule. It attracts other water molecules. When the molecules stick together, they are attached through Hydrogen Bonds, giving the molecule a property called **cohesion**. Cohesion gives water an unusually strong surface tension, and its capillary action makes the water spread. When the water spreads, **adhesion**, the tendency of water to stick to other materials, allows water to adhere to solids, making them wet.
- Water is the only known substance that readily exists in all three states of matter: liquid, solid, and gas. The hydrogen bonds holding water together are important. If they didn't exist, water would fly apart to form a gas. However, water primarily wants to be a liquid, and its liquid state range is meaned at 16°C.

Counting all sources, there is an impressive to compare the ratio of the mass of the ocean amount of water on the Earth:
98% of the water on Earth is in the oceans.
1.9% is locked up in glaciers as land ice.
0.5% is groundwater.
0.2% is found in rivers and lakes.
0.001% is present in the atmosphere.

It's very important to realize the scope and scale of water in comparison to the Earth's size: 61% of the Northern Hemisphere surface area and 81% of the Southern Hemisphere surface area is covered by water. The average land elevation is 840 meters, while the mean sea depth is 3,800 meters. If the Earth were completely covered by water, the land would be 2,440 meters (8,005 ft.) underwater. The total volume of water on Earth is estimated to equal in volume a cube 148 miles on each side. Unfortunately, the majority of water is undrinkable without treatment. In every kilogram of ocean water, 965 grams is freshwater, and 35 grams is inorganic salts. This non-potable factor is what leads to the noted water shortages in over 80 countries. In actuality, the Earth isn't short of water, but short of drinkable water, which comprises less than 0.7% of the total water available.

Calculating the Amount of Water on the Earth

To prove the fact that we have massive amounts of water on and within the Earth, we only need to compare the ratio of the mass of the ocean and the mass of the Earth's mantle. The Earth's mantle is made up of the stony meteorite mass from billions of years past. Water is locked within this iron mass. The percentage of mass of stony meteorites made up of water is 0.5 %.

Calculations

Mass of the Earth's mantle = density *volume.
$(4.5 g/cm^3) \times (1.0 \times 10^{27} cm^3) = 4.5 \times 10^{27}$ g of mantle mass.

Mass of the ocean:
$(1.025 g/cm^3) \times (1.4 \times 10^{24} cm^3) = 1.4 \times 10^{24}$ g mass of the ocean.

Percentage of mass of the mantle in ratio with the mass of the oceans:

$$\frac{\text{Ocean Mass}}{\text{Mantle Mass}} = \frac{1.4 \times 10^{24} g}{4.5 \times 10^{27} g} = 0.00031 = 0.031\%$$

Ratio of percent of meteorite mass composed of water to the minimum percentage of mantle mass:

$$\frac{\text{Meteorite}}{\text{Mantle}} = \frac{0.5}{0.31} = 16.1$$

There is 16.1 times more water in the mantle than currently in the oceans. The bottom line is that the mantle provided all the water for the oceans.

Ocean Facts
Surface area covered by 361,100,000 km
Volume of surface 1,370,000,000 km
Average depth of the ocean 3,796 meters
Average temperature of ocean 3.9 °C (39.0 °F)
Average salinity of ocean water 34.482 grams per kg
Most abundant elements in ocean water:
      Oxygen= 86%
      Hydrogen= 11 %
      Chlorine= 1.9%
      Sodium= 1.1 %
      Magnesium= O.1%

Dissolved Gasses in Seawater
Dissolved gasses in the seawater are very important to the marine life cycle. Surface gasses in the air easily dissolve in the ocean water. In terms of relative abundance, the major dissolved gasses are: Nitrogen, Oxygen, and Carbon Dioxide. Oxygen is the second most abundant gas in the ocean, with a content of around 36%. The third most abundant dissolved gas is Carbon Dioxide, comprising roughly 15% of the dissolved gasses in seawater.

The pH Balance
Maintaining a proper balance between acids and bases is critical to sustaining life in the oceans. This balance is measured on the pH scale, which measures the concentration of hydrogen ions in a solution. Pure water is considered neutral and is pegged at a pH of 7.0. As the numbers go lower, the acidity increases. As the numbers go higher, the alkaline (base) properties increase.

Acid: a substance that releases a hydrogen ion in solution. An acidic solution has an excess of (H+) ions.

Base: a substance that combines with a hydrogen ion in solution. A base (alkaline) solution has an excess of hydroxide (OH-) ions.

The increase or decrease from a neutral pH is based upon the interactions of the hydrogen and hydroxide ions in the water. Hydrogen ions (H+) and Hydroxide ions (OH-) are found in equal concentrations in pure water. Seawater is not pure water. It actually is slightly alkaline with a pH of 7.4 to 8.4. The median acidity is around 8.0-8.1. This alkalinity is maintained despite the high concentrations of Carbon Dioxide because of the forms the CO_2 takes in the seawater. Although the CO_2 does combine with the water to make Carbonic Acid (H_2CO_3), some of this acid breaks down to produce hydrogen ions (H+), bicarbonate ions (HCO_3-), and carbonate ions (CO_3^{2-}). This breakdown results in the maintenance of the alkalinity, because as the pH of the ocean drops (increases in acidity), a reaction occurs which removes more H+ ions, returning the water to the proper balance. Likewise, if the pH level increases (becomes more alkaline), more H+ ions are added to the water, once again maintaining the proper balance. This self-correcting feature of the seawater is called **buffering**.

Oceanic Depth and the pH Balance

Although the overall pH level of seawater is 8.0 to 8.1, the various water layers have different pH levels. Surface layers are generally warmer and have and abundance of marine plant life. As the plants use the CO_2, the pH level changes because the exchange process removes H+ ions. This results in a pH level of approximately 8.5. The middle and deep layers generally have more CO_2 present because of the lack of photosynthetic plant life (little sunlight penetration of these layers). The layers' depth drives a change in pressure and temperature, and the deeper you go, the more the pH balance shifts in an acidic direction. The water is still alkaline, but much less so than in the upper levels. The pH level below 15,000 meters is around 7.5. At the very lowest depths (below 18,000 ft), bacteria consumes oxygen and produces hydrogen sulfide that in turn, can lower the pH to that of pure water, 7.0.

SKILL 5.2   Describe the mechanisms that cause wave action and tides.

Waves

A transfer of energy from tides, currents, or wind causes waves. Water will remain in place unless moved by the current or tide. Currents are caused by changes in water density, salinity, and pressure. Tides are primarily caused by the gravitational pull of the moon. Waves move in an orbital pattern, causing an up and down motion. They have a forward or lateral motion only if moved by the wind, current, or tides. The depth of the waveform where the energy is felt is equal to ½ of the wavelength. Below that depth, the water remains relatively calm. When a wave approaches the shore, the circular orbit action flattens out and becomes more elliptical. As the wavelength shortens, the wave steepens until it finally breaks, creating surf. The waves break at a distance of 1/20 of the wavelength.

Tides

The periodic rise and fall of the liquid bodies on Earth are the direct result of the gravitational influence of the Moon and to a much lesser extent, the Sun. Tides are produced by the differences between gravitational forces acting on parts of an object. As shown in Netwon's Universal Law of Gravitation, the gravitational effect of two bodies is mutually constant and depends largely on the distance and mass between the objects. The side of the Earth that faces the Moon is roughly 4,000 miles (6,400 km) closer to the moon than is the Earth's center. This has the effect of increasing the Moon's gravitational attraction on Earth's oceans and landforms. Although the effect is so small on the mass of the landforms as to be invisible, the effect on the liquid parts is greater. The Moon's gravitational effect causes a bulge to form on both sides of the Earth. If we were able to view such subtle change from outer space, the affected waters would create an elliptical shape, compressing downward at the top and bottom of the planet and extending outward on the sides. This double-bulge effect causes the tides to fall and rise twice a day, and the time of the high and low tides is dependent on the phase of the moon. Yet not all locations are uniformly affected. The tidal cycle at a particular location is actually a very complicated interaction of the location's latitude, shape of the shore, etc. Because of its distance from the Earth, the Sun's gravitational effect on tides is only half that of the Moon's. However, when the gravitational effects of both the Sun and Moon join together during a new moon and a full moon phase, the tidal effects can be extreme. During a new moon and a full moon, tidal effects are much more pronounced as the tidal bulges join together to produce very high and very low tides. This pronounced type of tide are collectively known as Spring Tides. During the first and third quarters of the moon phases, the Sun's effect is negligible and consequently, the tides are lower. These are Neap Tides.

SKILL 5.3 Explain the layered structure of the oceans, including the generation of horizontal and vertical ocean currents and the geographic distribution of marine organisms, and how properties of ocean water, such as temperature and salinity, are related to these phenomena.

The ocean is commonly divided into five zones according to depth. Each zone has its own characteristics and a specific group of organisms reside there. The depth of water is key because photosynthesis requires solar energy. Therefore, the uppermost layer, where the light can penetrate, is where we can find phytoplankton. Phytoplankton are small, photosynthetic organisms which are the base of the oceanic food chain.

Epipelagic Zone – This layer extends from the surface to 200 meters (656 feet). It is in this zone that most of the visible light exists. The majority of plankton and fish are found here, as well as their predators (large fish, sharks, and rays).

Mesopelagic Zone - Extending from 200 meters (656 feet) to 1000 meters (3281 feet), the mesopelagic zone is also referred to as the twilight or midwater zone. Very little light penetrates here. Instead, most of the light observed is generated by bioluminescent creatures. A great diversity of strange fishes can be found here.

Bathypelagic Zone – This layer extends from 1000 meters (3281 feet) down to 4000 meters (13,124 feet). Here there is no penetration by solar light, so any light seen is in the form of bioluminescence. Most of the animals that live at these depths are black or red colored due to the lack of light. The water pressure at this depth is quite large, but a surprising number of creatures can be found here. Common inhabitants include fish, molluscs, jellies, and crustaceans. Sperm whales can dive down to this level in search of food.

Abyssopelagic Zone - Extending from 4000 meters (13,124 feet) to 6000 meters (19,686 feet), this zone has the least inhabitants. The water temperature is near freezing, and there is no light at all. Common organisms include invertebrates such as basket stars and squids. The name of this zone comes from the Greek meaning "no bottom", and refers to the ancient belief that the open ocean was bottomless.

Hadalpelagic Zone - This layer extends from 6000 meters (19,686 feet) to 10,000 meters (32,810 feet)- the sea floor. These areas are most often found in deep water trenches and canyons. In spite of the unimaginable pressures and cold temperatures, life can be found here. Generally, these include life forms that tolerate cool temperatures and low oxygen levels, such as starfish and tubeworms. The exception to this rule would be chemosynthetic communities living near deep-sea vents. These creatures create their own nutrients from carbon dioxide or methane released by the hot thermal vents. Chemosynthetic organisms then become prey to larger organisms. As such, chemosynthetic organisms are also primary producers and are at the bottom of the food chain, just like their photosynthetic friends, although they are at the opposite end of the ocean!

Layers of the Ocean

The layers of the ocean can be compared to the layers of an onion. Like an onion, the various layers overlap each and meet at distinctive boundaries

All the different layers of circulation are eventually tied together at the CDW (Circumpolar Deep Water), which has its origin in the North Atlantic.

Underneath the South Equatorial Current is a Subsurface Equatorial Counter-Current that flows eastward. Shaped much like a flat ribbon, the subsurface counter-current is only 200 meters thick and 2-3 degrees wide and it is located right on the geographical equator.

In general the Deep Water Layer flows south, while the Antarctic Bottom Water layer flows north.

The Pacific water is much more uniform in temperature at depth. Likewise, the Salinity Index in the Pacific is very uniform. The Pacific waters take a long time to circulate in comparison to the Atlantic.

The Atlantic Ocean waters are much younger than that of the Pacific. The margin of overturn is greater than two times that of the Pacific. It's estimated that the Atlantic waters overturn every 600 years. The Pacific is on the order of 1500-1600 years.

The North Atlantic Deep & Bottom waters are major factors in ocean circulation because of the effect of the Gulf Stream. The Gulf Stream has higher salinity and density and a huge flow south after looping around Iceland. However, to really determine where the various layers begin and end, and how far they travel, we need to use Temperature-Salinity (TS) charts. To determine the fine, small changes, we must look at salinity.

Salinity and Density

Salinity: the measure of the total concentration or amount of dissolved inorganic solids in water. Chloride (18%) and Sodium (10%) are the most abundant solids in the water.

The mean temperature of the ocean is 3.5 °C.

The mean salinity is 35 ppt (parts per thousand).

Salinity changes the physical properties of water in four significant ways. Collectively, these four changes are referred to as water's Colligative Properties.

- Heat capacity of water decreases with a rise in salinity. Less heat is necessary to raise the temperature of seawater by 1° than is required to raise the temperature of freshwater by an equal amount.
- Hydrogen bonding is disrupted by the concentration of salts in the water. The freezing point of seawater is lower than that of freshwater.
- Seawater evaporates more slowly than freshwater. The saline components tend to attract water molecules, causing the seawater to linger in the same circumstances under which freshwater evaporates.
- The Osmotic pressure exerted on a biological membrane is different from that within the cells when salinity increases.

At 5,500 meters depth, salinity remains extremely constant, and the current moves slowly with very little motion. There is a bigger density change in warmer water. The North Atlantic is saltier than the South Atlantic. Water sinks until it reaches its potential density. Example: The waters off of the Straits of Gibraltar on the ocean side have almost no slope and the bottom is all rock. This is because of the cascade of high saline, fast moving water out of the Mediterranean Sea. The South Atlantic and the Indian Ocean central waters are almost identical. This is due to their connection via the Circumpolar Central Current underwater. The Indian Ocean salinity is driven up by the outflow of the Red Sea, which is a Mediterranean Type Net Evaporation Marginal Sea. The Indian Ocean South Central waters mix with the Western South Pacific Central waters.

COMPETENCY 6.0 UNDERSTAND THE ATMOSPHERE.

SKILL 6.1 Compare the layers of the atmosphere in terms of chemical composition and thermal structure.

The Atmosphere

The Earth's atmosphere is very similar to a fluid. The atmosphere makes up only 0.25% of what we call the Earth, and like the fluids of the ocean, our atmosphere is driven by heat, primarily solar radiation. Having an atmosphere is not unique for a planet. To a degree, most of the planets in our solar system have an atmosphere, however, the presence of significant oxygen in Earth's atmosphere is unique and makes life possible on our planet. Earth's atmosphere is composed of 78% Nitrogen, 21% Oxygen, and 1% other gasses.

Layers of the Atmosphere

Troposphere: (Ground level to 11Km, 0 to 17.6 miles, or 0 to 92,928 feet.) The Troposphere varies in height according to the temperature. It is lower at the poles and higher at the equator. Because the pressure decreases, it gets colder as you go up in the Troposphere. Only very rarely do you have a mixing between the Troposphere and the next layer, the Stratosphere. All storms, weather fronts, and weather occur in the Troposphere.

Stratosphere: (11Km to 50Km, 17.6 to 80 miles, or 92,980 to 422,400 feet.) The Stratosphere is characterized by weak vertical air motion, and strong horizontal air motion. There is very little lifting or sinking air in the Stratosphere. Temperatures warm as you go up due to the presence of the Ozone layer contained within the Stratosphere.

Mesosphere: (50 to 85Km, 80 to 136 miles, or 422,400 to 718,080 feet.) It is bitterly cold in the Mesosphere.

Thermosphere: (85 to 600Km, 136 to 960 miles, or 718,000 to 5,068,800 feet.) This is the hottest portion of the atmosphere with rapid warming accompanying a rise in altitude. There are very few molecules left to block out the incoming solar radiation. The outer reaches of the Thermosphere are also sometimes referred to as the Exosphere.

Ionosphere: (Located within the upper portion of the Mesosphere at 80Km and goes into the Thermosphere). The ionosphere is an area of **free ions**: positively charged ions, produced as a result of solar radiation striking the atmosphere. The solar wind strikes the Ionosphere at the polar dips in the Magnetosphere. The ions are excited to a higher energy state and this energy is released into the visible spectrum to form the Aurora Borealis (Northern Lights). The Ionosphere varies with the time of day, season, and Sunspot cycles: When the Sunsets at night, less ions strike, extending radio wave communications. There is more radiation during Sunspot cycles. These hypercharge the atmosphere and can disrupt radio waves during the daytime.

SKILL 6.2 **Discuss the evolution of Earth's atmosphere over geologic time, including the effects of outgassing, the variations of carbon dioxide concentration, and the origin of atmospheric oxygen.**

Earth's **initial atmosphere** was composed of primarily hydrogen and smaller amounts of helium. However, most of the hydrogen and helium escaped into space very shortly after the earth was formed, approximately 4.6 billion years ago. A **second atmosphere** formed during the first 500 million years of Earth's history, as the gasses trapped within the planet were out- gassed during volcanic eruptions. This atmosphere was composed of carbon dioxide (CO_2), Nitrogen (N), and water vapor (H_2O), with smaller amounts of methane (CH_4), ammonia (NH_3), hydrogen (H), and carbon monoxide (CO). However, only trace quantities of oxygen were present. At around 3.5 billion years, Earth's **third atmosphere** began to form as the first life forms- simple, unicellular bacteria- appeared. As the primitive life forms evolved, they gradually developed the ability to conduct photosynthesis. **Photosynthsis** is the ability to combine Carbon dioxide and water in the presence of sunlight to for glucose and oxygen. The oldest evidence of life (3.4-3.5 billion years ago) is found in the Proterozoic Eon. All life during that period was protozoan, a microbial life form. The organisms used the glucose for food and released the oxygen into the atmosphere. as the early life forms thrived and multiplied over the next 3 billion years, they gradually released increasingly greater quantities of oxygen into the atmosphere. At around 2.5 billion years ago, the **Oxygen Revolution** took place, marking the point at which sufficient oxygen had accumulated to prompt an explosive evolutionary step in life formation. More complicated- but still unicellular- organisms began to appear. At around 570 million years ago the atmosphere reached the present day ration of approximately 78% nitrogen and 21 % oxygen. This shift marked another turning point in evolution as the oxygen level has reached a point to sustain more evolved life forms, and multicellular plants and animals made their first appearance.

TEACHER CERTIFICATION STUDY GUIDE

SKILL 6.3 Know the location of the ozone layer in the upper atmosphere, explain its role in absorbing ultraviolet radiation, and explain the way in which this layer varies both naturally and in response to human activities.

Ozone Layer (O_3): (Contained within the Stratosphere). Ozone is essential to life on Earth and is continually formed and destroyed within the atmosphere. Only a very thin layer of ozone protects against UV (ultra violet) radiation. Ultra violet radiation scrambles the DNA codes in human cells, and can kill the cells or, at a minimum, cause cancer. Much concern is made in the press about a hole in the ozone layer. This is a misnomer. In reality, there is not a hole, but a possible thinning of the layer. Many scientists believe this is due to the presence of CFC's (Carbon Fluorocarbons). The chlorine (Cl) in CFCs and from other sources steals an oxygen atom from Ozone (O_3) molecules. This theft leaves behind only plain oxygen (O_2). This does not effectively screen out uv radiation. $Cl + O_3 = ClO + O_2$. However, the resultant ClO molecule is very unstable and uv radiation can easily break it apart. The released chlorine then attacks another ozone molecule and the process repeats itself.

CFCs were thought to be the primary culprit, but this theory has problems. CFCs were only invented in the 1920's for use in aerosol spray cans, industrial processes, and refrigerants. Although it may be a contributor in the depletion of the ozone layer, not all scientists believe the theory that it CFC's are solely responsible. Why does the thinning occur over only the Antarctica? The greatest use of CFCs occurred in the industrialized nations of the Northern Hemisphere. Additionally, the hole varies in size from year to year, appearing during the Antarctic spring in October and disappearing by mid November or December.

There is a lack of data. Data was collected on this thinning only since 1979. There is a possibility that the hole may have been there before the introduction of CFCs. An alternate theory to CFCs is **Circumpolar Vortex**. Because there is a great deal of Open Ocean at Antarctica, it is a very cold place during the winter. The cold, in effect, isolates the Antarctic atmosphere from the rest of the warmer atmosphere. This extreme cold forms ice crystals in the atmosphere and Chlorine (Cl) is locked into the crystals. When the Antarctic spring comes, the atmosphere thaws out, releasing the chlorine into the atmosphere. The chlorine attacks the ozone layer, decreasing its density. As spring progresses, the circumpolar vortex weakens, allowing the air to mix with the normally ozone rich air. Thus, the hole disappears.

SKILL 6.4 Identify the bands at specific latitudes where rainforests and deserts are distributed and the causes of this pattern.

The location of a desert is a direct result of global air circulatory patterns. The equator receives more solar radiation. As the heated air rises, it moves both north and southward where it cools and sinks near the 30 degrees North and South latitudes. The great deserts of the world lay between these boundaries. These conditions exist primarily because the air, as it sinks back down from the atmosphere, compresses and is able to hold more water vapor. This in turn causes the water evaporated off of the surface to seldom fall back to the surface as rain. The opposite conditions exist at the Equator, where rainfall is abundant (rainforests).

COMPETENCY 7.0 UNDERSTAND EARTH'S ENERGY BUDGET INCLUDING INFLOW AND OUTFLOW.

SKILL 7.1 Compare the amount of incoming solar energy, the Earth's internal energy, the energy used by society, and the energy reflected back to space.

The total amount of incoming energy to the Earth's atmosphere is estimated at 1.74×10^{17} Watts. Incoming solar radiation accounts for 99.978 % of the total flux of energy entering the Earth's atmosphere. Other sources of incoming energy include geothermal energy, tidal energy and waste heat from fossil fuel consumption.

Internal geothermal energy accounts for 2.3×10^{13} W of the incoming energy to Earth's atmosphere. This energy is produced when stored heat or radioactive decay within Earth's interior rises to warm volumes of water beneath the Earth's surface. These volumes of water are known as geothermal deposits.

Worldwide energy consumption by the human race was measured in 2004 as 15×10^{12} Watts. 86.3 % of this energy use came from the burning of fossil fuels. Other sources of energy used by society include nuclear power and sources of renewable energy such as hydropower, solar power, wind power and geothermal power.

Outgoing energy from the Earth consists almost entirely of reflected solar radiation. The contributions of geothermal and tidal energy sources to the outgoing energy flux are negligible. 30 % of incoming solar radiation is reflected back into space, while 70 % is absorbed by Earth and eventually reradiated. Of the 30 % of solar radiation that is reflected by Earth, 6 % is reflected by the atmosphere, 20 % reflected by clouds and 4 % is reflected by land, water and ice. Of the 70% of solar energy absorbed by the Earth, 64 % is reradiated by the clouds and atmosphere, and 6 % is reradiated by the ground.

Energy recycled in the Earth's atmosphere is responsible for warming the Earth's surface. This process is known as the greenhouse effect. When in equilibrium, the total amount of energy entering the Earth's system from solar radiation will be exactly equal to the total amount of outgoing energy.

SKILL 7.2 Describe what happens to incoming solar radiation as it relates to reflection, absorption, and photosynthesis.

Radiation Budget: the balanced exchange cycle of radiation absorbed and released by the Earth's surface, water and atmosphere.

The Sun drives a radiation exchange interaction with the Earth every day as the Sun's radiated energy enters the Earth's sphere of influence. 30 % of incoming solar radiation is reflected back into space, while 70 % is absorbed by Earth and eventually reradiated. Of the 30 % of solar radiation that is reflected by Earth, 6 % is reflected by the atmosphere, 20 % reflected by clouds and 4 % is reflected by land, water and ice. Of the 70% of solar energy absorbed by the Earth, 64 % is reradiated by the clouds and atmosphere, and 6 % is reradiated by the ground.

The radiation absorbed by CO_2, dust, clouds, and water vapor in the atmosphere traps the heat and holds it to moderate the temperature (especially at night). In the desert, there is little water vapor present to trap the heat. That's why it's so cold at night. The interaction between the Earth's upper atmosphere, waters, and land keeps the world at a moderate ambient temperature of approximately 55 degrees F.

The distribution of solar energy is called **insolation**. Solar radiation isn't distributed evenly across the Earth, because of the Earth's curvature, axial tilt, and orbit. This results in uneven heating of the atmosphere, and is why the temperature is warmer at the equator and colder at the poles. Because of the curvature and tilt, the energy striking the polar areas is spread over a larger area. At the equator it is more concentrated. The same amount of energy is striking the atmosphere, but it is striking a larger or smaller area. It in effect, this dilutes the energy received by a particular area. The effect of insolation is very important to life on Earth. The absence of solar radiation would cause the creation of very cold air masses and the thermal blanket of the atmosphere would not have heat to hold and reradiate. In short order, the world would become an icy rock.

Plants have the ability to make their own food. Green organelles called Chloroplasts are located in the cytoplasm. These chloroplasts contain chlorophyll, a greenish pigment that traps light energy and through chemical reaction, changes it into chemical energy stored in sugar molecules. This process is known as **Photosynthesis**.

SKILL 7.3 Explain the mechanism and evaluate the significance of the greenhouse effect.

The Greenhouse Effect

The Greenhouse Effect is predicated on a rise in the trapped gases in the atmosphere.

CO_2, Methane, and Water Vapor all absorb reflected heat. CO_2 makes up 0.03% of the atmosphere. If the CO_2 level rises, more reflected heat is trapped, and the balance of the **Radiation Budget** is upset. When greenhouse gases and heat build up, the Earth's surface and atmospheric temperature rises. The theory contends that if we cut the amount of rising CO_2 in the atmosphere, then things will cool down.

Sources of greenhouse gases:
Gasoline burning engines and Aerosols (CO_2)
Vegetation & Microwaves devices (CO_2)
Rice Paddies & Ruminant animals (Methane)

CO_2 levels have risen this century, with the most dramatic rise since the 1960's. However, the data is not conclusive when compared to the past. The only thing certain is that there is a one-to-one correlation between temperature and CO_2.

The possibility of the oceans acting as a major source of CO_2 release cannot be discounted. Warm liquids hold less gas. As liquids cool, the gas goes into solution. Therefore, a change in the ocean temperatures either releases or holds more gases.

SKILL 7.4 Differentiate among greenhouse conditions on Earth, Mars, and Venus; the origins of those conditions and the climatic consequences of each.

Venus

In many respects, Venus is the twin of Earth. Based solely on size, its topography is suggestive of our own complex, geologic history. However, in terms of density and atmosphere, Venus is dramatically different from the Earth. Many cosmologists rank Venus as the most inhospitable planet in the entire solar system. Based on the information garnered from exploratory probes that have landed on the planet, we know that it is drier than any desert on Earth. The intense temperatures produced by Venus' atmosphere causes this extreme dryness.

The entire planet is covered in thick atmospheric clouds extending to a height of 68 km from the surface. These clouds are very stable despite wind speeds of 240 km/h and are carried rapidly around the planet by jet streams formed from convection currents rising from the surface. The dense cloud cover retains the heat in a **Runaway Greenhouse Effect**, causing the average surface temperature of 472 °C. The extremely dense atmosphere is primarily composed of 95% carbon dioxide (CO_2), and 3.5% nitrogen (N). The remaining percentage is divided among sulfuric acid (H_2SO_4), hydrochloric acid (HCl), hydrofluoric acid (HF), and trace amounts of water.

Because of its proximity to the Sun, if Venus ever had any significant quantity of surface water, it would have soon evaporated. Without oceans to regulate the amount of CO_2 retained in the atmosphere, heat is trapped inside of the atmosphere. This retained heat eventually dehydrated the surface materials. The CO_2 is produced by the high degree of volcanism on the planet. Massive shield volcanoes form titanic volcanic mountains, thick highlands, and extensive lava plains on the surface of Venus.
Example: The Ishtar Terra highlands are larger than the United States, and the volcanic mountain, Maxwell Montes is 12 km high.

Earth

Earth is the third planet outward from the Sun. As the Earth is already detailed in other parts of this book, the information presented in this section will be brief. The atmosphere is composed of nitrogen and oxygen. Earth is unique among all the planets of the solar system for the extent of water present. Over 70% of the planet is composed of water. The presence of the water makes possible the protective atmosphere and more important, life forms. This planet is the only known location of life in the universe. To become a watery planet, specific criterion must be met:

- An ocean world must move in a nearly circular orbit around a stable star and the distance of the planet from the star must be just right to provide a temperature environment in which water is liquid.
- A watery planet's sun must not be a double or multiple star, or the orbital year would have irregular periods of intense heat and cold.
- The materials that accreted to form the planet must include both water and substances capable of forming a solid crust. Likewise, volcanoes or steaming vents are needed to vent water vapor to the surface.
- The planet must have enough mass, density, and gravity to create and keep an atmosphere, preventing the oceans from drifting off in space.

Earth is also the only known planet to meet the additional conditions needed to support life.

- Earth's gravity is strong enough to retain an ocean, but not strong enough to crush the life forms that came from it.
- The planet has a magnetic field provided by the iron core that deflects radiation that would otherwise harm the genetic instructions of the organisms.
- The single moon causes relatively gentle tides that encourage life forms to leave the sea and reside on land.
- The atmosphere is relatively clear so that sunlight penetrates to the surface, but moist enough to form rains and winds that drive the ocean currents.
- The upper atmosphere contains Ozone (O_3), which protects against the most harmful ultraviolet (UV) rays.

Mars

Mars has a rotation and axial inclination very similar to Earth, 1 Martian Yr. = 2 Earth Years and 1 Martian day = Slightly over 24 Earth hours.

Geologically, Mars' composition is believed to be the almost identical to Earth's. There is significant evidence that surface water was once present on the Martian surface. There is currently no known water present and its ice caps are frozen carbon dioxide (dry ice). The very low density of the planet and thin atmosphere do not provide enough pressure to prevent liquid water from boiling into vapor. The Martian atmosphere is primarily carbon dioxide (CO_2) with trace amounts of oxygen, argon and other gasses. However, the atmosphere is thick enough to create wind effects. Large, wind formed dune fields mark the surface. Strong winds and large dust storms are still prevalent. Mars is a cold planet. The surface temperature ranges between -140 °C to 20 °C.

COMPETENCY 8.0 UNDERSTAND CIRCULATION IN THE OCEANS AND ATMOSPHERE.

SKILL 8.1 Assess the differential effects of heating on circulation patterns in the atmosphere and oceans.

Ocean circulation: The direction indicated by the name is going to that direction. Example: The Northwest current is going to the northwest.

Convergence: where warm and cold wafers are pinched together. The denser water will sink below the lighter density water. A convergence zone implies sinking. It also means a horizontal change in temperature and salinity, and thus, a change in density.

Geostrophic Flow

Pressure gradient and Coriolis effect based, the Geostrophic flow model states that current flow parallel to a hill can be created in the ocean by changes in density, pressure, and temperature. A massive horizontal change in temperature also changes the density and pressure. In example, the Gulf Stream has one of the biggest horizontal pressure changes in the ocean. By plotting the temperature and salinity gradients, you can develop an idea of the pressure gradient. This allows you to determine the direction of flow of the current.

SKILL 8.2 Relate the rotation of Earth to the circular motions of ocean currents and air in low- and high-pressure centers.

Water moves in a curved path because of the rotation of the Earth. The **Coriolis Effect** is the deflection of air or water currents caused by the rotation of the Earth. This creates global wind patterns that affect the climate. These wind patterns also represent rain patterns. The wind patterns between 30°N and 0°N are called the **Trade Winds**. The wind patterns between 30°N and 60°N are called the **Prevailing Westerlies**.

SKILL 8.3 Compare the causes and structures of various cloud types, precipitation, air masses, and fronts, and the causes and effects of different types of severe weather.

Clouds are classified by their physical appearance and given special Latin names corresponding to the cloud's appearance and the altitude where they occur. Classification by appearance results in three simple categories: cirrus, stratus, and cumulus clouds. Cirrus clouds appear fibrous. Stratus clouds appear layered. Cumulus clouds appear as heaps or puffs, similar to cotton balls in a pile. Classification by altitude result in four groupings: high, middle, low, and clouds that show vertical development. Other adjectives are added to the names of the clouds to show specific characteristics.

Cloud Classifications

High Clouds: -13 °F (-25 °C) Composed almost exclusively of ice crystals
<u>Cirrus</u> >23,000 ft (7,000 m). Nearly transparent, delicate silky strands (mare's tails), or patches.
<u>Cirrostratus</u> >23,000 ft (7,000 m). A thin veil or sheet that partially or totally covers the sky. Nearly transparent, the sun or moon readily shines through.
<u>Cirrocumulus</u> >23,000 ft (7,000 m). Small, white, rounded patches arranged in a wave or spotted mackerel pattern

Middle Clouds: 32 - -13 °F (0 - -25 °C) Composed of supercooled water droplets or a mixture of droplets and ice crystals
<u>Altostratus</u> 6600 - 23,000 ft (2000 - 7000 m). Uniform white or bluish-gray layers that partially or totally obscure the sky layer
<u>Altocumulus</u> 6600 - 23,000 ft (2000 - 7000 m). Roll-like puffs or patches that form into parallel bands or waves

Low Clouds: > 23 °F (-5 °C) Composed mostly of water droplets
<u>Stratocumulus</u> 0-6,600 ft (0-2000 m). Large irregularly shaped puffs or rolls separated by bands of clear sky
<u>Stratus</u> 0-6600 ft (0-2000 m). Uniform gray layer that stretches from horizon to horizon. Drizzle may fall from the cloud
<u>Nimbostratus</u> 0-13,120 ft (0-4000 m). Thick, uniform gray layer from which precipitation (significant rain or snow) is falling

Clouds with Vertical Development: Water droplets build upward and spread laterally
<u>Cumulus</u> 0-9840 ft (0-3000 m). Resemble cotton balls dotting the sky.
<u>Cumulonimbus</u> 0-9840 ft (0-3000 m). Often associated with thunderstorms, these large puffy, clouds have smooth or flattened tops, and can produce heavy rain and thunder.

All forms of precipitation start from an interaction of water vapor and other particulate matter in the atmosphere. These particulates act as a nucleus for raindrops, as the water vapor particles attached themselves to the other airborne particles. Because one of water's major properties is that its water particles attract other water particles, the raindrop grows as water vapor particles accrete around the nuclei.

Drizzle: any form of liquid precipitation where the drops are less than 0.02 inches in diameter.

Rain: any form of liquid precipitation where the drops are greater than 0.02 inches in diameter.

Virga: the meteorological condition where rain evaporates before touching the ground. You see it rain, but it never hits the ground.

Snow: water molecules that form into ice crystals through freezing. The shape of the snowflakes depend on the temperature at which they formed:

Needles = 0°C to -10°C.
Dendrites= -10°C to -20°C.
Plates= -20°C to -30°C.
Columns= -30°C to -40°C.

Freezing Rain: drops fall as rain but immediately depose (freeze) upon hitting an extremely cold surface such as power lines, roofs, or the ground. This is also called an **Ice Storm**.

Rime Ice: ice droplets that have tiny air bubbles trapped within the ice, producing an opaque whitish layer of granular ice.

Sleet: officially called ice pellets, these are drops of rain 5mm or less in diameter. Sleet freezes before hitting the ground and bounce when they strike a surface.

Hail: precipitation in the form of balls or lumps of ice. Hail forms when an ice pellet is transported through a cloud that contains varied concentrations of super cooled water droplets. The pellet may descend slowly through the entire cloud, or it may be caught in a cycle of updraft and downdraft. The ice pellet grows by accreting (adding) freezing water droplets. Eventually, weight of the hail grows too heavy to be supported by the air column and falls to the ground as a hailstone. The size of the stone depends on the amount of time spent in the cloud.

Knowledge of types of storms

A **thunderstorm** is a brief, local storm produced by the rapid upward movement of warm, moist air within a cumulo-nimbus cloud. Thunderstorms always produce lightning and thunder, accompanied by strong wind gusts and heavy rain or hail. **Lightning**: a brilliant flash of light produced by an electrical discharge of about 100 million volts. Lightning flashes when the attraction between positive and negative charges (ions) becomes strong enough to overcome the air's normally high resistance to electrical flow. Normally, the surface of the Earth is negatively charged and the upper troposphere is positively charged. However, this distribution changes when a cumulonimbus cloud develops. Charges separate within the cloud so that the upper portion and a small region near the base become positively charged. Likewise, the cloud induces a positive charge on the ground directly beneath it. As a thunderstorm matures, electrical resistance of the air breaks down and lightning can flow either between oppositely charged areas of the cloud, or between the cloud and the ground.

Tornado: an area of extreme low pressure, with rapidly rotating winds beneath a cumulonimbus cloud. Tornadoes are normally spawned from a Super Cell Thunderstorm. They can occur when very cold air and very warm air meet, usually in the Spring. Worldwide, the U.S. has the most tornadoes and most of these occur in the spring. Texas has the most tornadoes, but Florida has the largest number per square mile. Tornadoes are without a doubt the most violent of all storms. Roughly 120 people each year are killed in the United States by tornadoes. Authorities may issue a **Tornado Watch** if meteorological conditions could/ probably will cause the formation of mesocyclones. A **Tornado Warning** is issued when a fume cloud is spotted.

A swirling, funnel-shaped cloud that **extends** downward and touches a body of water is called a **waterspout.**

Hurricanes are produced by temperature and pressure differentials between the tropical seas and the atmosphere. Powered by heat from the sea, they are steered by the easterly trade winds and the temperate Westerlies, as well as their own incredible energy. Hurricane development starts in June in the Atlantic, Caribbean, and Gulf of Mexico, and lasts until the end of hurricane season in late November. Hurricanes are called by different names depending on their location. In the Indian Ocean they are called **Cyclones**. In the Atlantic, and east of the international dateline in the Pacific, they are called Hurricanes. In the western Pacific they are called **Typhoons**. Regardless of the name, a hurricane can be up to 500 miles across, last for over two weeks from inception to death, and can produce devastation on an immense scale. The destruction and damage caused by a hurricane or tropical storm can be severe. **Storm surge** causes most of the damage as the winds push along a wall of rising water in their path, and this rising effect is amplified on low sloping shorelines such as found on the Gulf Coast. The intense winds can also cause damage.

Storms that occur only in the winter are known as blizzards or ice storms. A **blizzard** is a storm with strong winds, blowing snow and frigid temperatures. An **ice storm** consists of falling rain that freezes when it strikes the ground, covering everything with a layer of ice.

Types of Fronts

Front: a narrow zone of transition between air masses of different densities that is usually due to temperature contrasts. Because they are associated with temperature, fronts are usually referred to as either warm or cold.

Warm Front: a front whose movement causes the warm air (less dense) to advance, while the cold air (more dense) retreats. A warm front usually triggers a cloud development sequence of cirrus, cirrostratus, altostratus, nimbostratus, and stratus. It may result in an onset of light rain or snowfall immediately ahead of the front, which gives way as the cloud sequence forms, to steady precipitation (light to moderate), until the front passes, a time frame that may exceed 24 hours. The gentle rains associated with a warm front are normally welcomed by farmers. However, if it is cold enough for snow to fall, the snow may significantly accumulate. If the air is unstable, cumulonimbus clouds may develop, and brief, intense thunderstorms may punctuate the otherwise gentler rain or snowfall.

Cold Front: a front whose movement causes the cold air (more dense) to displace the warm air (less dense). The results of cold front situations depend on the stability of the air. If the air is stable, nimbostratus and altostratus clouds may form, and brief showers may immediately precede the front. If the air is unstable, there is greater uplift, cumulonimbus clouds may tower over nimbostratus and cirrus clouds are blown downstream from the cumulonimbus by the winds at high altitude. Thunderstorms may occur, accompanied by gusty surface winds and hail, as well as other, more violent weather. If the cold front moves quickly (roughly 28 mph or greater), a squall line of thunderstorms may form either right ahead of the front or up to 180 miles ahead of it.

Occluded Front: a front where a cold front has caught up to a warm front and has intermingled, usually by sliding under the warmer air. Cold fronts generally move faster than warm fronts and occasionally overrun slower moving warm fronts. The weather ahead of an occluded front is similar to that of a warm front during its advance, but switches to that of a cold front as the cold front passes through.

Stationary Front: a front that shows no overall movement. The weather produced by this front can vary widely and depends on the amount of moisture present and the relative motions of the air pockets along the front. Most of the precipitation falls on the cold side of the front.

SKILL 8.4 Know and explain features of the ENSO cycle (El Niño southern oscillation, including La Niña) in terms of sea-surface and air temperature variations across the Pacific, and climatic results of this cycle.

El Nino is a reverse of the normal weather patterns in the Pacific. A low-pressure area normally sits in the Pacific Ocean west of Hawaii and a high-pressure area normally sits off of the California coast. When an El Nino forms, these pressure areas shift eastward, causing the low-pressure area to be situated below Hawaii and the high-pressure area to move inland over California. Because of the shift in pressure areas, an El Nino affects the wind patterns (especially the jet stream and trade winds), and creates a wide variety of effects., including a direct impact on commercial fishing. Normally, there is a shallow warm water layer over the colder, deeper waters along the coastlines. The temperature disparity causes an upwelling of rich nutrients from the lower layers of the cold water, creating a feeding zone that attracts a variety of marine life forms. In an El Nino situation, the warm water layer increases in both area coverage and depth. It extends downward, blocking the nutrient rich upwell from reaching the feeding zone. Although many species can migrate to more friendly waters, some have limited mobility and die. In any event, the fishery area can become permanently barren depending on the intensity, duration, and repetition of El Nino events. Other effects of El Nino are both direct and indirect.

Direct Effects
- In the west, fires and drought.
- In the east, rain, landslides, and fish migration.

Indirect Effects
- Greater chance of hurricanes in Hawaii.
- Lesser chance of hurricanes in Virginia because of weakened Trade Winds.
- Less rain during September and October.
- Coastal erosion in the western states.
- Fewer storms that deposit snow in the Cascades in Washington and Oregon.
- The Jet Streams are altered as the high pressure areas move.

A **La Nina** is the opposite of El Nino. However, it does not have as great effect. It causes the East to be wetter and the west to be drier. Many scientists believe that both of these conditions are caused by a change in the surface temperature of the water of the Pacific Ocean.

COMPETENCY 9.0 UNDERSTAND CLIMATE VARIATIONS IN TIME AND SPACE.

SKILL 9.1 Analyze weather (short-term) and climate (over time) in relation to the transfer of energy into and out of the atmosphere.

Climate: the characteristic weather of a region over long periods of time.

Weather: the state of the atmosphere at a particular time and place.

Climate Zones

The movement of the air forms **Convection Currents** that affect the weather and climate in three zones. The patterns created in these zones distribute the heat between the poles and the equator.

The **Coriolis Effect** is the deflection of air or water currents caused by the rotation of the Earth. This creates global wind patterns that affect the climate. These wind patterns also represent rain patterns. The wind patterns between 30°N and 0°N are called the **Trade Winds**. The wind patterns between 30°N and 60°N are called the **Prevailing Westerlies**. All the great deserts of the world lie between the Tropics at 0° and 30°North & South latitudes. A shift in the wind patterns would also shift the deserts. The optimum growing zone is between 30° and 60° North & South latitudes.

If global warming takes place, it could cause the melting of the polar ice caps that in turn could raise sea levels. This would inundate land areas and lead to economic disaster.

The Earth's climate is in a very delicate balance. A change to the global warming or cooling patterns would affect retained heat, which would further affect the growing zones. In example, global warming would cause the expansion of the tropics, while global cooling would cause a contraction of the tropic zone.

SKILL 9.2 Discuss and assess factors that affect climate including latitude, elevation, topography, and proximity to large bodies of water and cold or warm ocean currents.

A number of factors affect the climate of a given location. These factors include latitude, elevation, topography, and proximity to large bodies of water and cold or warm ocean currents.

Latitude

Latitude is the measure of how far you are north or south of the equator. Latitude has a pronounced effect on climate because it determines how much solar energy a location receives. Locations close to the equator receive more direct sun light and, thus, more radiant energy and heat. Locations distant from the equator (and nearer the Earth's poles) receive less direct sun light as the sun's rays strike the Earth at an angle. Thus, locations closer to the poles receive less radiant energy and heat.

Locations between the Tropics of Capricorn and Cancer, located approximately 23 degrees south and north of the equator respectively, experience a tropical climate, typified by hot, moist weather throughout the year. The Arctic and Antarctic Circles, located at approximately 66 degrees north and south of the equator respectively, define the Polar Regions. Locations north of the Arctic Circle and south of the Antarctic Circle experience a polar climate, typified by extremely cold temperatures and varying lengths of days. Finally, the temperate zones, defined as the areas between the tropics and the Polar Regions, experience seasonal climates, typified by cold weather in the winter and warm weather in the summer.

Elevation

In general, elevation, or distance above sea level, has a clear effect on climate. When comparing two locations at the same latitude, the one with the higher elevation will have a cooler climate. As you go up a mountain, for instance, the air pressure decreases and the air gets less dense. Less dense air does not hold heat as well as more dense air does. Thus, locations at higher elevations have consistently cooler climates.

Topography

Topography, or the features of the land, can also affect a location's climate. For example, the presence or absence of mountains can greatly affect the amount and location of precipitation in a given region. Mountains typically receive a large amount of precipitation because less dense air at higher elevations cannot hold as much moisture. As air rises up a mountain precipitation is more likely to occur. Conversely, as air currents travel over a mountain and down the other side, the air heats up, becomes denser, and is able to hold more moisture. Thus, the leeward side of mountains (or side facing away from the wind) is often very dry. For example, Death Valley in California is a leeward desert caused by the aforementioned mountain effect.

Proximity to Large Bodies of Water

Large bodies of water have a pronounced affect on the climate of surrounding landmasses, by moderating temperatures and increasing precipitation. Coastal areas generally experience cooler temperatures during the summer and warmer temperatures during the winter than their inland counterparts. This moderation of temperature results from the ability of large bodies of water to absorb a large amount of solar energy. During the summer, bodies of water trap heat, keeping temperatures cooler. Conversely, during the winter, the latent heat stored in the water escapes to warm the atmosphere slightly. Finally, coastal areas generally receive more precipitation because the bodies of water serve as a direct source of moisture.

Cold or Warm Ocean Currents

There is a direct relationship between the temperature of ocean currents and the temperature of surrounding landmasses. Coastal areas near warm ocean currents experience warmer climates than coastal areas at similar latitudes near cold ocean currents. Warm ocean currents heat the air above the ocean, while cold ocean currents cool the air above the ocean. The varying atmospheric temperatures and related wind patterns act to warm or cool the related coastal regions.

Milankovitch Cycles

This theory proposes that the axial tilt and wobble of the Earth's orbit is responsible for warming or cooling.

Precession: the wobble of the Earth on its axis.

Obliquity: the tilt of the Earth's axis. This ranges between 22.5 to 25.5 degrees. Today the tilt is 23.5 degrees.

Eccentricity: the shape of an orbit. The Earth's orbital shape changes periodically from elliptical to circular.

The theory is based on the Earth's relationship to the Sun.

If there is a change in the tilt and orbit, then the amount of incoming solar radiation will change over time.

COMPETENCY 10.0 UNDERSTAND THE ROCK CYCLE.

SKILL 10.1 Compare and contrast the properties of rocks based on physical and chemical conditions in which rocks are formed, including plate tectonic processes.

The **rock cycle** is a dynamic process of ongoing change that reflects the recycling of the Earth's materials. Although the processes involved in the rock cycle are dynamic, they follow a geological, rather than human time scale, and consequently, changes occur so slowly that they are not readily observable.

Weathering: the physical and chemical breakdown and alteration of rocks and minerals at or near the Earth's surface.

Mechanical (Also called Physical Weathering: where rock is broken into smaller pieces with no change in chemical or mineralogical composition. The resulting material still resembles the original material.
Example: Rock pieces breaking off a boulder.
The pieces still resemble the original material, but on a smaller scale.

Chemical Weathering: where a chemical or mineralogical change occurs in the rock and the resulting material no longer resembles the original material.
Example: Granite (Gneiss/Schist) eventually weathers into separate sand, silt, and clay particles.

Weathering is typically caused by a combination of chemical and mechanical processes.

Factors Influencing Weathering:
- **Composition:** Due to their composition, some rocks weather easier and will show more effects.
- **Rock Structure:** Does it have cracks? Is it fractured? Water and other elements get in the cracks.
- **Climate:** The more water, the more the weathering effect. Additionally, the higher the temperature or the more the temperature varies, the greater the weathering effect.
- **Topography:** This factor determines the amount of surface area exposed to weathering. Smaller rocks are affected more because collectively, they have less mass and more surface area than a boulder.
- **Vegetation:** Important weathering agent. Depending on the type of vegetation, it can either hinder or accelerate the weathering process. Although vegetation may leave less surface area exposed, the vegetation's root structures can produce a biological effect that accelerates the process.

Types of Mechanical Weathering:
- **Frost Wedging:** This occurs when rock gets a crack in it, water collects, and then freezes. Over time, as this cycle repeats itself, the expanding water gradually pushes the rock apart.
- **Salt Crystal Growth:** In a process similar to frost wedging, as the water evaporates, it leaves salt crystals behind. Eventually, these crystals build up an push the rock apart. This is a very small-scale effect and takes considerably longer than frost wedging to affect the rock material.
- **Abrasion:** This is a key factor in mechanical weathering. The motion of the landscape materials produces significant weathering effects, scouring, chipping, or wearing away pieces of material. Abrasive agents include wind blown sand, water movement, and the materials in landslides bashing into each other.
- **Biological Activity:** This is a two-fold weathering agent.
    - **Plants:** Seeds will sometimes land in a crack in a rock and begin to grow in the cracks. The root structure eventually acts as a wedge, pushing the rock apart.
    - **Animal:** As animals burrow, the displaced material has an abrasive effect on the surrounding rock. Because of the limited number of burrowing animals, plant activity has a much greater weathering effect.
- **Pressure Release (Exfoliation):** Rock expands when compressive forces are removed, and bits of the rock break off during expansion. This can result in massive rock formations with rounded edges.
- **Thermal Expansion and Contraction:** Minerals within a rock will expand or contract due to changes in temperature. Dependent on the minerals in the rock, this expansion and contraction occurs at different rates and to different magnitudes. Essentially, the rock internally tears itself apart. The rock may look solid but when placed under pressure, easily crumbles.

Climate is a key factor in mechanical weathering.

Types of Chemical Weathering:
- **Oxidation (Rust):** Oxygen atoms become incorporated into the formula of a mineral in a rock and the mineral becomes unstable and breaks off in flakes.
- **Solution:** Due to their inherent composition, some minerals found in rocks easily dissolve into solution when exposed to a liquid.
- **Acids:** Water and water vapor may combine with other elements and gases to form acids. Water (H_2O) and carbon dioxide (CO_2) can chemically combine to become Carbonic Acid (H_2CO_3). Sulfur Dioxide (SO_2) particles can chemically combine with water (H_2O) to form Sulfuric Acid (H_2SO_4).
- **Biological Activity:** Plant roots growing in the cracks of rocks not only cause mechanical wedging, but also secrete acids that cause chemical weathering.

Generally found in combination with solution, acids cause the majority of chemical weathering.

SKILL 10.2 Identify common rock-forming minerals (e.g., feldspars, quartz, biotite, calcite) using a table of diagnostic properties.

Biotite (Mica Group): The iron is what gives the Biotite its dark color as compared to another mineral, Muscovite, which doesn't have iron in it and has a transparent appearance similar to cellophane.

Quartz: Inert to most effects, quartz stays in the environment as original material.

Feldspar: forms clay minerals. Feldspar is the most common rock-forming mineral and comprises about 60% of the earth's crust. These can be identified as light colored minerals that break with a smooth surface. Feldspar minerals are usually white or very light in color, have a hardness of 6 on the Mohs' Scale of Hardness, and perfect to good cleavage in two directions. Ex. = Albite, $NaAlSi_3O_8$.

Calcite: The chemical formula for calcite is $CaCO_3$. It constitutes the majority of sedimentary and metamorphic limestone.

Albite	Colorless, lustrous, perfect or good cleavage, uneven fractures with flat surfaces
Biotite Biotite (continued)	Black color, one perfect direction of cleavage resulting in the mineral pealing into thin, flexible sheets, similar properties to Muscovite
Calcite	H=3, reacts with HCl, 3 directions of cleavage (rhombic cleavage)
Quartz	H=7, conchoidal fracture, no cleavage, color is typically white or clear but can be pink, red, purple, black

SKILL 10.3 Identify common ore minerals as sources of copper, iron, lead, zinc, cement, halite, gypsum, and uranium.

Mineral resources are concentrations of naturally occurring inorganic elements and compounds located in the Earth's crust that are extracted through mining for human use. Minerals have a definite chemical composition and are stable over a range of temperatures and pressures. Construction and manufacturing rely heavily on metals and industrial mineral resources. These metals may include iron, bronze, lead, zinc, nickel, copper, tin, etc. **Orebites** are economically important ore minerals. Other industrial minerals are divided into two categories: bulk rocks and ore minerals. Bulk rocks, including limestone, clay, shale and sandstone, are used as aggregate in construction, in ceramics or in concrete. Common ore minerals include calcite, barite and gypsum. Energy from some minerals can be utilized to produce electricity fuel and industrial materials. Mineral resources are also used as fertilizers and pesticides in the industrial context.

COMPETENCY 11.0 UNDERSTAND THE WATER, CARBON, AND NITROGEN CYCLES.

SKILL 11.1 Illustrate the mechanism that drives the water cycle.

The hydrologic cycle of water movement is driven by solar radiation from the Sun. The cycle of evaporation from the oceans, and precipitation over land is the methodology employed by nature to maintain the water balance at any given location. The Earth constantly cycles water. It evaporates from the sea, falls as rain, and flows over the land as it returns to the ocean. The constant circulation of water among sea, land, and the atmosphere is called the **hydrologic cycle**.

Evaporation
Water is constantly in motion on the Earth. As the water evaporates from the sea, it becomes water vapor in the atmosphere. Although a small amount of water evaporates from the land and inland waterways, the majority of evaporation occurs over the oceans. An additional small amount of evaporated water comes from plants as they breathe using the process of **transpiration**.

Precipitation
The water vapor in the atmosphere is returned to the Earth in the form of precipitation. Precipitation includes rain, hail, snow, and sleet. The amount of precipitation varies according to location, with some areas of Earth receiving plentiful moisture, and others receiving little (the deserts). However the overall proportional balance of evaporation and precipitation remains relatively constant.

Runoff
As the moisture returns to the Earth's surface in the form of precipitation, the liquid moves across the land according to the topology, with most of the water eventually flowing back into the oceans. Thus the cycle starts over: evaporation, precipitation, and runoff.

SKILL 11.2 Compare the processes of photosynthesis and respiration in terms of reservoirs of carbon and oxygen.

Cellular respiration is the metabolic pathway in which food (glucose, etc.) is broken down to produce energy in the form of ATP. Both plants and animals utilize respiration to create energy for metabolism. In respiration, energy is released by the transfer of electrons in a process know as an **oxidation-reduction (redox)** reaction. The oxidation phase of this reaction is the loss of an electron and the reduction phase is the gain of an electron. Redox reactions are important for the stages of respiration.

Glycolysis is the first step in respiration. It occurs in the cytoplasm of the cell and does not require oxygen. Each of the ten stages of glycolysis is catalyzed by a specific enzyme. Beginning with pyruvate, which was the end product of glycolysis, the following steps occur before entering the **Krebs cycle**.

1. Pyruvic acid is changed to acetyl-CoA (coenzyme A). This is a three carbon pyruvic acid molecule which has lost one molecule of carbon dioxide (CO_2) to become a two carbon acetyl group. Pyruvic acid loses a hydrogen to NAD^+ which is reduced to NADH.

2. Acetyl CoA enters the Krebs cycle. For each molecule of glucose it started with, two molecules of Acetyl CoA enter the Krebs cycle (one for each molecule of pyruvic acid formed in glycolysis).

The **Krebs cycle** (also known as the citric acid cycle), occurs in four major steps. First, the two-carbon acetyl CoA combines with a four-carbon molecule to form a six-carbon molecule of citric acid. Next, two carbons are lost as carbon dioxide (CO_2) and a four-carbon molecule is formed to become available to join with CoA to form citric acid again. Since we started with two molecules of CoA, two turns of the Krebs cycle are necessary to process the original molecule of glucose. In the third step, eight hydrogen atoms are released and picked up by FAD and NAD (vitamins and electron carriers).
Lastly, for each molecule of CoA (remember there were two to start with) you get:

>3 molecules of NADH x 2 cycles
>1 molecule of $FADH_2$ x 2 cycles
>1 molecule of ATP x 2 cycles

Therefore, this completes the breakdown of glucose. At this point, a total of four molecules of ATP have been made- two from glycolysis and one from each of the two turns of the Krebs cycle. Six molecules of carbon dioxide have been released; two prior to entering the Krebs cycle, and two for each of the two turns of the Krebs cycle. Twelve carrier molecules have been made- ten NADH and two $FADH_2$. These carrier molecules will carry electrons to the electron transport chain. ATP is made by substrate level phosphorylation in the Krebs cycle. Notice that the Krebs cycle in itself does not produce much ATP, but functions mostly in the transfer of electrons to be used in the electron transport chain where the most ATP is made.

In the **Electron Transport Chain,** NADH transfers electrons from glycolysis and the Kreb's cycle to the first molecule in the chain of molecules embedded in the inner membrane of the mitochondrion. Most of the molecules in the electron transport chain are proteins. Nonprotein molecules are also part of the chain and are essential for the catalytic functions of certain enzymes. The electron transport chain does not make ATP directly. Instead, it breaks up a large free energy drop into a more manageable amount. The chain uses electrons to pump H^+ across the mitochondrion membrane. The H^+ gradient is used to form ATP synthesis in a process called **chemiosmosis** (oxidative phosphorylation). ATP synthetase and energy generated by the movement of hydrogen ions coming off of NADH and $FADH_2$ builds ATP from ADP on the inner membrane of the mitochondria. Each NADH yields three molecules of ATP (10 x 3) and each $FADH_2$ yields two molecules of ATP (2 x 2). Thus, the electron transport chain and oxidative phosphorylation produces 34 ATP.

So, the net gain from the whole process of respiration is 36 molecules of ATP:

Glycolysis - 4 ATP made, 2 ATP spent = net gain of 2 ATP
Acetyl CoA- 2 ATP used
Krebs cycle - 1 ATP made for each turn of the cycle = net gain of 2 ATP
Electron transport chain - 34 ATP gained

Photosynthesis is an anabolic process that stores energy in the form of a three carbon sugar. We will use glucose as an example for this section. Photosynthesis is done only by organisms that contain chloroplasts (plants, some bacteria, some protists). There are a few terms to be familiar with when discussing photosynthesis.

An **autotroph** (self feeder) is an organism that make its own food from the energy of the sun or other elements. Autotrophs include:

1. **photoautotrophs** - make food from light and carbon dioxide releasing oxygen that can be used for respiration.
2. **chemoautotrophs** - oxidize sulfur and ammonia; this is done by some bacteria.

Heterotrophs (other feeder) are organisms that must eat other living things for their energy. **Consumers** are the same as heterotroph; all animals are heterotrophs. **Decomposers** break down once living things. Bacteria and fungi are examples of decomposers. **Scavengers** eat dead things. Examples of scavengers are bacteria, fungi and some animals.

The **chloroplast** is the site of photosynthesis. It is similar to the mitochondria due to the increased surface area of the thylakoid membrane. It also contains a fluid called stroma between the stacks of thylakoids. The thylakoid membrane contains pigments (chlorophyll) that are capable of capturing light energy.

Photosynthesis reverses the electron flow. Water is split by the chloroplast into hydrogen and oxygen. The oxygen is given off as a waste product as carbon dioxide is reduced to sugar (glucose). This requires the input of energy, which comes from the sun.

Photosynthesis occurs in two stages: the light reactions and the Calvin cycle (dark reactions). The conversion of solar energy to chemical energy occurs in the light reactions. Electrons are transferred by the absorption of light by chlorophyll and cause the water to split, releasing oxygen as a waste product. The chemical energy that is created in the light reaction is in the form of NADPH. ATP is also produced by a process called photophosphorylation. These forms of energy are produced in the thylakoids and are used in the Calvin cycle to produce sugar.

The second stage of photosynthesis is the **Calvin cycle**. Carbon dioxide in the air is incorporated into organic molecules already in the chloroplast. The NADPH produced in the light reaction is used as reducing power for the reduction of the carbon to carbohydrate. ATP from the light reaction is also needed to convert carbon dioxide to carbohydrate (sugar).

The process of photosynthesis is made possible by the presence of the sun. Visible light ranges in wavelengths of 750 nanometers (red light) to 380 nanometers (violet light). As wavelength decreases, the amount of energy available increases. Light is carried as photons, which is a fixed quantity of energy. Light is reflected (what we see), transmitted, or absorbed (what the plant uses). The plant's pigments capture light of specific wavelengths. Remember that the light that is reflected is what we see as color. Plant pigments include:

Chlorophyll *a* - reflects green/blue light; absorbs red light
Chlorophyll *b* - reflects yellow/green light; absorbs red light
Carotenoids - reflects yellow/orange; absorbs violet/blue

The pigments absorb photons. The energy from the light excites electrons in the chlorophyll that jump to orbitals with more potential energy and reach an "excited" or unstable state.

The formula for photosynthesis is:

$$CO_2 + H_2O + \text{energy (from sunlight)} \rightarrow C_6H_{12}O_6 + O_2$$

The high energy electrons are trapped by primary electron acceptors which are located on the thylakoid membrane. These electron acceptors and the pigments form reaction centers called photosystems that are capable of capturing light energy. Photosystems contain a reaction-center chlorophyll that releases an electron to the primary electron acceptor. This transfer is the first step of the light reactions. There are two photosystems, named according to their date of discovery, not their order of occurrence.

The production of ATP is termed **photophosphorylation** due to the use of light.

Below is a diagram of the relationship between cellular respiration and photosynthesis.

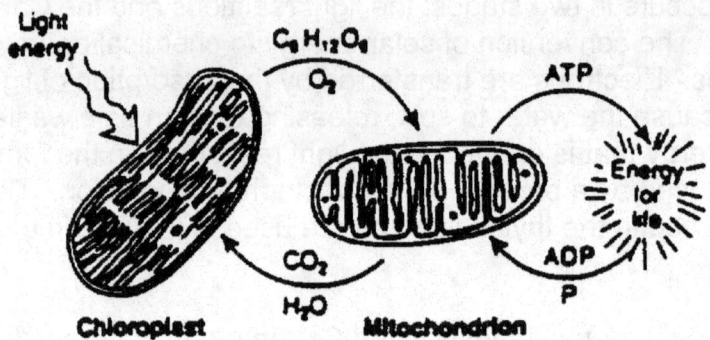

SKILL 11.3 Identify the carbon reservoirs (i.e., physical and chemical forms of carbon in the atmosphere, oceans, biomass, soils, fossil fuels, and solid earth) and describe the movement of carbon among these reservoirs in the global carbon cycle.

Life on Earth is made up of a variety of carbon molecules that are often called the building blocks of life. In essence, life processes are carbon chemistry. The carbon cycle can be thought of as the movement of carbon atoms through various storage places (reservoirs) on Earth. In addition, the water cycle and carbon cycle are intricately related. The water cycle helps to drive the carbon cycle, and this is where climate and the carbon cycle are most closely connected.

Great carbon reservoirs exist in the crust of the Earth. These include limestone rocks and the coal deposits (fossil fuels). Nitrate and phosphate reservoirs can be found in the ocean. The atmosphere s itself a reservoir. It contains 78% Nitrogen, 21% Oxygen, and 1% other gasses. There is increasing concern of greenhouse gasses, including Carbon dioxide. Carbon dioxide is a chemical combination of carbon and oxygen (CO_2). It is a byproduct of human exhalation as well as pollution. The reservoirs have interfaces were they connect. Organic matter decays in soils and the resulting gases are released into the air. Oxygen is incorporated into the ocean and utilized according to depth. Nutrients are eroded from or leached out of the soil and transported into the sea by runoff.

The carbon cycle describes the movement of carbon atoms through the systems of the planet. Examples of carbon reservoirs include the ocean, the atmosphere, the biosphere, the soil, carbonate sediments found at the bottom of the ocean and also organic carbon matter. The quantifiable changes between the carbon reservoirs attest to the rate at which atoms move from one reservoir into another.

Physical Carbon Pump

The exchange of carbon between the atmosphere and ocean takes place in multiple ways. One of these mechanisms is through physical mixing of the ocean (vertical deep mixing). It occurs when warm water in oceanic surface currents are carried from low to high latitudes and cooled, making them heavy. They then sink below the surface layer and in some places all the way to the bottom. This is why the deep ocean is significantly colder than the surface temperature. When seawater is cooled it can take up more carbon dioxide. When cold water returns to the surface it warms again and loses carbon dioxide to the atmosphere. Vertical circulation ensures that carbon dioxide is constantly moving between the ocean and the atmosphere. Thus vertical circulation is a carbon pump.

The Biological Carbon Pump

The ocean gets a disproportionate share of the carbon dioxide available to the ocean-atmosphere system. The ratio is about 50 molecules of CO_2 in the ocean for every one in the atmosphere. This is because carbon dioxide readily reacts with water to make soluble bicarbonate (HCO_3-). This cold, carbon dioxide-rich water is then pumped down by vertical mixing into lower depths. Another reason for the ocean's large share of carbon is biological. The biological pump removes carbon dioxide from the ocean's water by converting it into living matter (organisms). This living matter is distributed among the water layers and not so easily accessible to the atmosphere.

Marine Carbon (Carbonate) Cycle

Both the mixing of the ocean and the biological processes described above are important with regard to the carbon budget of the sea and exchange with the atmosphere. Carbon is also stored in sediments and recycled.

The marine carbon cycle involves the production and recycling of two types of carbon-rich materials: organic matter and carbonate ($CaCO_3$). The production of solid $CaCO_3$ occurs in the surface waters of the ocean by organisms that build their shells from $CaCO_3$ and through the inorganic chemical balance as defined by: $Ca_{2+} + 2HCO_3- \rightarrow CaCO_3 + CO_2 + H_2O$.

The Soil Cycle

In the carbon cycle, decomposers recycle the carbon accumulated in durable organic material that does not immediately proceed to the carbon cycle. Plant debris are deposited and buried in the soil and bacteria and fungi digest the deposited organic material. Global warming is expected to increase this process in areas that have been traditionally frozen and unavailable.

SKILL 11.4 Describe the nitrogen cycle as it relates to the atmosphere, soils as reservoirs, life processes, and pollution.

Earth's largest reservoir for Nitrogen is air. Roughly eighty percent of the atmosphere is in the form of nitrogen gas. Nitrogen must be fixed (taken out of its gaseous form) to be incorporated into an organism. Such fixed, or useable, compounds include nitrate ions (NO_3^-), ammonia (NH_3), and urea ($NH_2)2CO$. Every organism requires nitrogen to make proteins and nucleic acids, which are essential to life. Animals obtain their nitrogen compounds from plants (or animals that have fed on plants). Ammonification is the decomposition of organic nitrogen back to ammonia. This process in the nitrogen cycle is carried out by aerobic and anaerobic bacterial and fungal decomposers. Decomposers add nitrogen and phosphorous back to the soil by decomposing the excreted waste of animals. Only a few genera of bacteria have the correct enzymes to break the triple bond between nitrogen atoms in a process called nitrogen fixation. These bacteria live within the roots of legumes (peas, beans, alfalfa) and add nitrogen to the soil so it may be taken up by the plant. Interestingly, nitrogen can also be 'fixed' by lightening production. The enormous energy of lightning breaks nitrogen molecules, enabling their atoms to combine with oxygen in the air forming nitrogen oxides. These dissolve in rain, forming nitrates, which are carried to the earth. Atmospheric nitrogen fixation probably contributes some 5–8% of the total nitrogen fixed.

COMPETENCY 12.0 UNDERSTAND TECTONIC EVOLUTION.

SKILL 12.1 Interpret geologic maps as a basis for understanding the tectonic evolution of California in terms of plate margins (i.e., Atlantic-type passive margin, Japanese volcanic arc, Andean arc, and faulted margin).

Evolution can be defined as change over time. When we speak of evolution, it is usually in terms of a species, but topography can change too. California is well known for its earthquakes. These earthquakes occur because of California's unique positioning.

Plate margins are labeled as destructive, conservative, collision, or constructive. A destructive margin is where an oceanic plate moves towards and is subducted under a continental plate. This situation causes volcanoes, earthquakes, and is responsible for folded mountains. Conservative margins occur where two plates slide past each other. This is true of the Juan de Fuca and North American plates, which cause many earthquakes in California. Collision margins are where two continental plates move toward each other. Because they are within continents, they create mountain ranges, such as the Himalayas. Constructive margins are present where two plate move away from one another.

Plate movements and the basic difference in the density of oceanic and continental crustal units initiate the structural pattern of continental margins and result in a tectonic classification of coastlines as active (Pacific, leading edge) or passive (Atlantic, trailing edge) margins, each of which have certain fundamental characteristics.

Active continental margins (leading edge margins) are defined as having converging plates. They coincide with plate boundaries, where the continental and oceanic crusts are separated by a subduction zone. These margins are tectonically active and have less width and sediment input than passive margins. Blocks from distant sources may also be added to the continental mass at the subduction zone.

Passive margins (also called trailing edge) are located within plates and are separated from the oceanic ridge plate margin by an expanse of oceanic crust that was generated after rifting. Oceanic and continental crust meet in a region of low tectonic activity. Passive margins are generally wide and may receive a large influx of sediments or carbonate sedimentation from local sources. The Atlantic margin is a passive margin.

The area between two plates sliding horizontally past one another is called a transform-fault boundary, or transform boundary. These large faults connect two spreading centers (or, less commonly, trenches). Most transform faults are found on the ocean floor and offset areas of spreading, thus producing zig-zag plate margins, and are generally defined by shallow earthquakes. However, a few do occur on land, for example the San Andreas fault zone in California. This transform fault connects the East Pacific Rise, a divergent boundary to the south, with the South Gorda/ Juan de Fuca/ Explorer Ridge, another divergent boundary to the north.

Subduction processes in oceanic-oceanic plate convergence result in the formation of volcanoes. Over millions of years, the erupted lava and volcanic debris pile up on the ocean floor until a submarine volcano rises above sea level to form an island volcano. Such volcanoes are typically strung out in chains called island arcs. Magmas that form island arcs are produced by the partial melting of the descending plate and/or the overlying oceanic lithosphere. The descending plate also provides a source of stress as the two plates interact, leading to frequent moderate to strong earthquakes.

Off the coast of South America along the Peru-Chile trench, the oceanic Nazca Plate is pushing into and being subducted under the continental part of the South American Plate. In turn, the overriding South American Plate is being lifted up, creating the towering Andes mountains, the backbone of the continent. Strong, destructive earthquakes and the rapid uplift of mountain ranges are common in this region. The Puyehue volcano in southern Chile is part of the Andean arc.

Japan's volcanoes are part of five volcanic arcs. These five arcs meet at the island of Honshu and form part of the 'Ring of Fire". The Northeast Honshu Arc and the Kurile Arc formed by the subduction of the Pacific Plate under the Eurasian Plate. The Izu-Bonin Arc is the result of subduction of the Pacific Plate under the Philippine Plate. The Southwest Honshu Arc and the Ryukyu Arc formed by the subduction of the Philippine Plate under the Eurasian Plate. Black "teeth" mark the subduction zone with the "teeth" on the overriding plate.

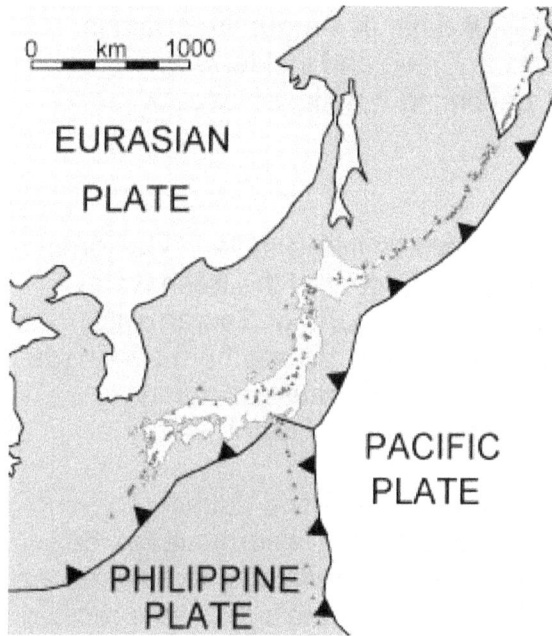

COMPETENCY 13.0 IDENTIFY MAJOR ECONOMIC EARTH RESOURCES.

SKILL 13.1 Understand the importance of water to society, the origins of California's fresh water, statewide water distribution, and the environmental and economic impact of water redistribution.

Water is essential to all living things. There were major problems with water pollution in the past. In 1969, a river running through Cleveland, Ohio, was so badly polluted that it caught fire and burned for 8 days. It caught fire again in 1979. Due to the international publicity from the Cleveland incident, water pollution problems received national focus, becoming a national issue.

Origins of California's Fresh Water

Historically, one of the biggest environmental problems in the state of California is water supply and distribution. Approximately 75 percent of the fresh water supply originates in the northern one-third of the state (north of Sacramento) while 80 percent of the demand for fresh water occurs in the southern two-thirds of the state. This imbalance in supply and demand is the cause of much disagreement between residents of different regions within California. [1]

The origins of California's fresh water include rivers, lakes, and deltas. The Sacramento-San Joaquin Delta is a unique and important California resource. The Delta receives runoff from 40 percent of the state's landmass and is the focal point of water distribution in the state. Two-thirds of the state's residents receive at least a portion of their drinking water from the Delta. Other important sources of water include the Owens River, the Colorado River, and Mono Lake. These three bodies of water account for a significant portion of the water supply to Los Angeles. The water supply to Los Angeles has traditionally been extremely problematic because of the dry climate of Southern California. [2]

Statewide Water Distribution

An extensive system of reservoirs, aqueducts, dams, and pumping stations distributes fresh water throughout the state of California, often moving water up to hundred's of miles from its original source. As previously discussed, the most daunting problem facing engineers is the need to deliver the water supply from the northern part of the state, to the more populous southern part of the state. Dams and reservoirs help concentrate water for storage and distribution. Aqueducts and pumps physically move the water from place to place. For example, a series of dams and aqueducts delivers water from the distant Colorado River to the cities of San Diego and Los Angeles.

---

[1] http://www.lib.berkeley.edu/WRCA/exhibit.html
[2] http://www.lib.berkeley.edu/WRCA/exhibit.html

Environmental and Economic Impact of Water Redistribution

As you might expect, the construction of dams, aqueducts, and associated water conveyance infrastructure is extremely costly. The California government must set aside a significant portion of the state budget for development and maintenance of the water distribution system.

Of equal importance is the strain water redistribution can place on the environment. Damming and removal of water from rivers and other bodies of water disrupts the natural environment. Diminished water levels can negatively affect local plants and animals. The instability of water levels regularly endangers many species of plants and animals. In addition, dams increase the likelihood of flooding. Unexpected weather conditions or dam structural problems can cause flooding, which can destroy property and cause death and destruction to humans, animals, plants, and associated habitats.

SKILL 13.2 Describe resources of major economic importance in California and their relation to California's geology (e.g., oil, gas, gold, sand, gravel, salts, open space, soil, arable land, clean air).

Because of its large size and diverse topography and climate, California has many natural resources of economic importance. These resources include oil, gas, gold, sand, gravel, salts, open space, soil, arable land, and clean air.

Oil

You may be surprised to learn that one of the chief exports of Southern California is oil.[3] Southern California hosts numerous underground oil reserves that oil companies have been exploiting over the past 100 years. These reserves, first indicated by visible oil seeps above ground, make California the fourth leading producer of oil in the United States (behind Louisiana, Texas, and Alaska).[4] Tectonic plate activity (e.g. the San Andreas Fault) that is responsible for seismic activity in California also creates a subsurface environment conducive for the production of oil.

---

[3] http://www.priweb.org/ed/pgws/history/signal_hill/signal_hill.html
[4] http://www.energy.ca.gov/oil/index.html

Natural Gas

Another important energy resource found in California is natural gas. Californians use natural gas, both domestic and imported, for electricity generation and for residential, industrial, and commercial purposes. Natural gas originates in two forms, associated and non-associated. Associated gas is produced along with crude oil, while non-associated gas derives from gas fields where crude oil is not produced. The bulk of the associated gas production occurs in Southern California while the bulk of the non-associated gas production occurs in Northern California. Similar to oil production, natural gas originates deep within the Earth, as a result of the formation and movement of certain types of rock.

Gold

Famous for the Gold Rush of 1848 at Sutter's Mill, California has long been known for its production of gold. A highly valuable commodity because of its unique properties and consumer appeal, gold is one of the most important natural resources in the history of California. Many regions in California produce gold, but the most productive districts are the northern and central portions of the Sierra Nevada.[5] Like oil and natural gas, gold results largely from subsurface plate activity, volcanic activity, and rock movement and formation.

Sand and Gravel

California is one of the leading producers of construction-grade sand and gravel. Sand and gravel are rounded rock and mineral fragments classified by the size of the particles. Sand and gravel are eroded products of rock formations. Scientists believe the movement of glaciers during the Ice Age is largely responsible for inland accumulations of sand and gravel. In addition, California has an extremely long coastline, where large depositions of sand are available for mining and use.

Salts

California also produces a significant amount of salt. The preferred method of salt production in California is evaporation. Several sites across the state host solar evaporation ponds filled with salt water. Over time, the sun evaporates the water leaving behind salt deposits for easy collection.

---

[5] http://www.pioneermining.com/dist_gold_dist.htm

Open Space

Open space is a precious commodity in any developed society. As the population of California continues to grow, there is increasing pressure to develop land previously used for agriculture or other open space purposes. In an attempt to curtail such developmental sprawl the California government adopted the California Land Conservation Act of 1965 (commonly known as the Williamson Act). This legislation allows local governments to enter into agreements with private land owners with the purpose of restricting development on certain parcels of open space land. In exchange for giving up their developmental rights, land owners receive significantly lower property tax assessments.[6]

Soil and Arable Land

Due to its favorable and diverse climate, California is a leading supplier of a wide range of agricultural products for a large portion of the United States. Many regions of California have high quality soil that is conducive to agriculture of many types. Soil conservation and replenishment is an important task that farmers must practice, especially because of California's soil characteristics. Because California soil is warm, regularly irrigated, and tilled, it does not favor the build up of organic materials. Thus, farmers must regularly replenish the soil to keep it arable and favorable for agricultural practices.

Clean Air

Clean air is another important economic resource. California has long had a problem maintaining clean air in the face of the large amounts of pollution originating from the Los Angeles area. The state's sunny climate, mountainous topography (that traps pollution), and the state's residents' energy using habits all contribute to the pollution problem. While the air is cleaner now than it was several decades ago, California still has much work to do to maintain and improve air quality.[7]

[7] http://www.arb.ca.gov/html/brochure/arb.htm

---

[6] http://www.consrv.ca.gov/DLRP/lca/index.htm
[7] http://www.arb.ca.gov/html/brochure/arb.htm

TEACHER CERTIFICATION STUDY GUIDE

COMPETENCY 14.0 EXPLAIN SURFACE PROCESSES.

SKILL 14.1 Assess mechanisms by which tectonics, geologic structures (i.e., folds and faults), and rock properties influence surface properties (e.g., flow of water, differential erosion, uplift, subsidence).

Plates are rigid blocks of earth's crust and upper mantle. These rigid solid blocks make up the lithosphere. The earth's lithosphere is broken into nine large sections and several small ones. These moving slabs are called plates. The major plates are named after the continents they are "transporting."

The plates float on and move with a layer of hot, plastic-like rock in the upper mantle. Geologists believe that the heat currents circulating within the mantle cause this plastic zone of rock to slowly flow, carrying along the overlying crustal plates.

Movement of these crustal plates creates areas where the plates diverge as well as areas where the plates converge. A major area of divergence is located in the Mid-Atlantic. Currents of hot mantle rock rise and separate at this point of divergence creating new oceanic crust at the rate of 2 to 10 centimeters per year. Convergence is when the oceanic crust collides with either another oceanic plate or a continental plate. The oceanic crust sinks forming an enormous trench known as a subduction zone and generating volcanic activity. Portions of the lithosphere are dragged into the mantle. Then some of this material melts and volcanoes erupt. In time, series of volcanic islands are formed like the Aleutian Islands that are parallel to the trench.

Convergence also includes continent to continent plate collisions. When two plates slide past one another a transform fault is created. The crustal movement which is identified by plates sliding sideways past each other produces a plate boundary characterized by major faults that are capable of unleashing powerful earthquakes. The San Andreas Fault forms such a boundary between the Pacific Plate and the North American Plate.

A mountain is terrain that has been raised high above the surrounding landscape by volcanic action, or some form of tectonic plate collisions. This is known as uplift. The plate collisions could be intercontinental or ocean floor collisions with a continental crust (subduction). The physical composition of mountains would include igneous, metamorphic, or sedimentary rocks; some may have rock layers that are tilted or distorted by plate collision forces.

There are many different types of mountains. The physical attributes of a mountain range depends upon the angle at which plate movement thrust layers of rock to the surface. Many mountains (Adirondacks, Southern Rockies) were formed along high angle faults.

EARTH & PLANETARY SCIENCE

Folded mountains (Alps, Himalayas) are produced by the folding of rock layers during their formation. The Himalayas are the highest mountains in the world and contain Mount Everest which rises almost 9 km above sea level. The Himalayas were formed when India collided with Asia. The movement which created this collision is still in process at the rate of a few centimeters per year.

Fault-block mountains (Utah, Arizona, and New Mexico) are created when plate movement produces tension forces instead of compression forces. The area under tension produces normal faults and rock along these faults is displaced upward.

Dome mountains are formed as magma tries to push up through the crust but fails to break the surface. Dome mountains resemble a huge blister on the earth's surface.

Upwarped mountains (Black Hills of S.D.) are created in association with a broad arching of the crust. They can also be formed by rock thrust upward along high angle faults.

Volcanism is the term given to the movement of magma through the crust and its emergence as lava onto the earth's surface. Volcanic mountains are built up by successive deposits of volcanic materials.

The Ice Age began about 2 -3 million years ago. This age saw the advancement and retreat of glacial ice over millions of years. Theories relating to the origin of glacial activity include Plate Tectonics, where it can be demonstrated that some continental masses, now in temperate climates, were at one time blanketed by ice and snow. Another theory involves changes in the earth's orbit around the sun, changes in the angle of the earth's axis, and the wobbling of the earth's axis. Support for the validity of this theory has come from deep ocean research that indicates a correlation between climatic sensitive micro-organisms and the changes in the earth's orbital status.

There are two main types of glaciers: valley glaciers and continental glaciers. Erosion by valley glaciers is characteristic of U-shaped erosion. They produce sharp peaked mountains such as the Matterhorn in Switzerland. Erosion by continental glaciers often rides over mountains in their paths leaving smoothed, rounded mountains and ridges.

Plains are vast flat areas that make up about one half the total landscape of the United States. The Atlantic Coastal Plain is the exposed margin of the continental shelf that has emerged from the sea during the past several million years. It is a lowland that rests upon soft, loose sediments such as sand, silt, and clay. Low hills and valleys occur only when the plain has been slightly uplifted. Wide river valleys have cut through the plan producing low hills between river systems. Much of the area is covered by swamps and lakes.

The Great Plains is a young landscape underlain by sediments that have been eroded from the uplifted Rocky Mountains. Streams have deposited these materials in nearly horizontal layers during the last few million years. The plains elevation ranges from 1500 meters at its western boundary to 350 meters at its eastern boundary. Therefore, rivers that cross the plain all flow east. Most of the landforms present in this area are related to the drainage patterns developed by river systems. Many complex drainage patterns have been etched into the loose, soft sediments of the plain.

Stream patterns provide clues to rock resistance in an area. A rock of uniform resistance will produce a dendritic stream pattern. A dendritic pattern resembles a tree with branches, so this is a pattern of creeks flowing into streams which flow into tributary rivers and eventually into a large river. Radial patterns, streams flowing outwardly from a central location, form on the slopes of volcanoes. Trellis patterns develop in alternating resistant and weak layers of tilted sedimentary rock. And rectangular patterns are seen when the rock is jointed or fractured in a rectangular pattern.

SKILL 14.2 Discuss the factors controlling the influence of water in modifying the landscape.

Landscape Alterations by Groundwater

Groundwater is very effective in dissolving limestone, forming caves and sinkholes. Chemical leeching combined with groundwater produces Carbonic acid ($CO_2 + H_2O = H_2CO_3$), and this acid dissolves the limestone, riddling an area with underground holes.

- Cave: an empty space underground. The majority of caves are formed by carbonate dissolution. 95% are made of limestone.
- Speleotherm: a general term for a deposit in a cave.
    - Stalactite: mineral formations from the ceiling.
    - Stalagmite: mineral formation built up from the floor.
    - Column: mineral formations at the top and bottom of the cave that have joined together.
- Sinkhole: a collapsed cave. Often a void in the limestone will fill with water and during times of drought, the water recedes from the cave. The water supported the material about the cave and the unsupported weight of the cave ceiling collapses. Sinkholes are unpredictable in terms of when and where they will occur. They can be huge in scope-swallowing up entire city blocks-and often become small lakes dotting the landscape.
- Thermal Spring (Hot Spring): These are common in volcanic regions where an active magma chamber is at depth. This heat source causes nearby groundwater to become warm where the water table is exposed to the surface (a spring).

- Geyser: a thermal spring that erupts. The processes behind the eruption are very similar to those involved in boiling water in a teakettle. A constriction forms in the connected chambers of a spring. The water heats under pressure, turns to steam, and erupts with great force past the constriction. The ejected steam condenses and returns to a liquid state. The water draws back into its chambers and the process begins again. Since it takes awhile for the water to drain back and reheat, geysers often erupt on a determinable schedule.

Coastal Geomorphology
- Drowned River Valleys: (Cutting). Characterized by V-shaped cutting.
- Fjords: (Cutting by Glaciers). Characterized by U-shaped cutting.
- Drumlins: (Filling).
- River Mouth Deltas: (Filling).
- Volcanoes: (Lava cones).
- Faulting: (Tensional breaking). Tectonic lifting or dropping to change shape of coast.
- Barrier Islands: (Tend to be wave straightened.) Barrier Islands make up 80% of
- the U.S. East Coast. Examples: Spits, Baymouth Bars, and Tomboloes.
- Wave Eroded: (Both straightened and irregular). The straightness is formed due to ocean action. Examples: Wave straightened shorelines typify straightened. Retreating cliffs are typify irregular.
- Cuspate Foreland: (Current sculpted). Characterized by huge, gentle curves in the shoreline.

SKILL 14.3 Interpret the factors controlling erosion, deposition, and transport in surficial processes.

Shoreline Erosion: Erosion tends to straighten a shoreline.
- Abrasion: sandblasting effect caused by particles of sand or sediment.
- Hydraulic Action: Waves striking a cliff drives water into the cracks or crevices in the rock, compressing air into the cracks. Together, the air and water may dislodge rock fragments or even huge boulders.
- Currents: The longshore current and rip current scoop out material and carry it along the shoreline or out to sea.

Shoreline Erosional Features
- <u>Sea Cave</u>: Erosive effects gradually hollow out cliff faces where the water strikes. This erosion continues until the water can no longer reach the back of the cave.
- <u>Wave Cut Platform (Wave-Cut Terrace)</u>: Sometimes the face of the cliff will collapse after a sea cave is dug by erosion. This forms a plateau type platform that is smoothed by further erosion.

- Sea Arch: The erosive effects of the waves have the greatest effect against the material of a headland. The erosion literally digs sea caves in from both sides of the headland. Eventually, these caves join together, forming a sea arch.
- Sea Stack: Formed when the top of a sea arch collapses, leaving a vertical standing piece of the land. Can be quite huge, resembling a small island. Example: California coastline has thousands of sea stacks.

Coastline Deposition and Beaches

Deposition along the coastline affects mostly unconsolidated particles, sand-sized or smaller, and it occurs whenever the Longshore Current slows. The main feature of this deposition is the building of beaches.

Beach: a large accumulation of sand and sand sized particles along a body of water. A beach is in equilibrium with the distinct forces that create it. Unless changes occur to disturb the equilibrium, the beach will remain constant in composition and size. The beach actually extends outward to a depth of 30 feet. After 30 feet, the sand is lost to the ocean bottom. The 30-foot depth can be as much as a mile offshore. Beaches change with the season:

- Winter: Strong waves push the sand into offshore bars. The effect of the stronger wave action during the winter months creates a steeper slope for the beach.
- Summer: Gentle waves move the sand bars onto the shoreline. In the summer, as the material is re-deposited to the beach, and the slope is gentler. The chief agent of transportation along the coastline is the Longshore Current (Drift).

The Longshore Current runs parallel to the coastline, and there is literally a river of sand carried down the beach shoreline by the current. This is not necessarily negative, because the sand is mostly re-deposited further down the beach. Overall, the transportation effect merely moves the sand with no net loss or gain. Because beaches attempt to remain in equilibrium, as they adjust, black material can be found on some beaches. The material is not oil spill, but it is muck from the swampy area behind the original berms. The beach moved inland, exposing the dried muck from the bottom of the swampy area.

Desert Erosion

The majority of rainfall is sourced from evaporation of seawater. The greater the distance from the ocean, when taken in conjunction with the presence of mountains creating a rain shadow effect, can lead to inland deserts of great scope. Because deserts typically have sparse vegetation and the water cannot soak into the sun-baked ground, its runoff is rapid, and often creates a **Flash Flood**. These torrents erode the landscape and are laden with sediment that, as the floodwaters subside, coats the now dry streambed with a layer of sand and gravel. The rapid movement of the water also encourages a heavy downcutting effect, sculpting narrow canyons with vertical walls and graveled floors. The shortage of water slows the chemical weathering process to such a degree that minerals seldom break down into fine-grained clay products. Desert features usually look angular rather than rounded, due to the differentiated erosion effect on the materials present. In the desert, Limestone is much more resistant to erosion than in wetter areas. Most igneous and metamorphic rocks are also resistant, and shale is the least resistant, forming a gentler slope.

SKILL 14.4 Appraise desert environments in terms of water resource needs for habitation.

Desert: any region with low rainfall. A region is normally classified as a dry or arid climate if less than 25 centimeters (10 inches) of rain falls during the year. Deserts typically have little rain, high evaporation, plentiful sunshine, and clear skies. Technically, despite the presence of vast ice sheets, the continent of Antarctica is the driest region on Earth, as it averages less than 1 inch of rain per year. The location of a desert is a direct result of global air circulatory patterns. The equator receives more solar radiation. As the heated air rises, it moves both north and southward where it cools and sinks near the 30 degrees North and South latitudes. The great deserts of the world lay between these boundaries.

Desert Characteristics

Because few plants can tolerate such conditions the deserts typically have a forlorn, barren appearance. The plants that do manage to survive the rigorous climate in desert regions are typically salt-tolerant. They have extensive root systems and are widely spaced. Leaves are characteristically small to minimize water loss during transpiration and during much of the year they resemble dead twigs. However, when rain does fall on the region, these same plants will rapidly green and bloom.

The shortage of water also drives most of the topological processes within a desert. Most of the time the streambeds are dry. This lack of through-flowing streams coupled with **internal drainage**- flowing toward the interior basin instead of toward the sea- causes the surface of each basin to act as a local base level.

COMPETENCY 15.0 UNDERSTAND NATURAL HAZARDS.

SKILL 15.1 Analyze published geologic hazard maps of California and know how to use maps to identify evidence of geologic events of the past and to predict the likelihood of geologic changes in the future.

Look for hazard maps at http://education.usgs.gov/california/resources.html

The National Hazard Maps show the distribution of earthquake shaking levels that have a certain probability of occurring in the United States. These maps were created to provide the most accurate and detailed information possible to assist engineers in designing buildings, bridges, highways, and utilities that will withstand shaking from earthquakes in the United States. These maps are used to create and update the building codes that are now used by more than 20,000 cities, counties, and local governments to help establish construction requirements necessary to preserve public safety.

How to read a hazard map

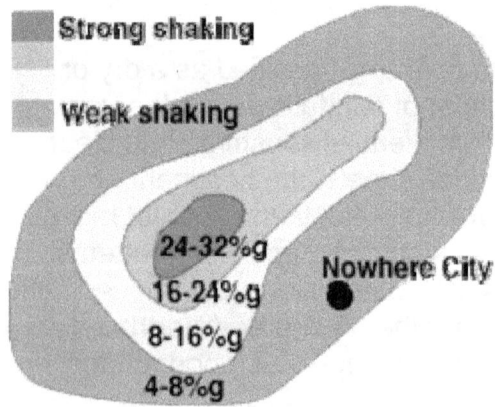

For the example map given:

1. A 50-year time interval
2. A 5% chance of exceedence
3. A PGA map

We would read the shaking hazards for Nowhere City as:

The earthquake peak ground acceleration (PGA) that has a 5% chance of being exceeded in 50 years has a value between 4 and 8% g.

There are 3 types of maps:

- **PGA** - Peak Ground Acceleration
- **SA 0.2 sec** - 0.2 second Spectral Acceleration
- **SA 1.0 sec** - 1.0 second Spectral Acceleration

Units for all 3 maps are %g (percent of gravity)

This can also be expressed in decimal form, example 10%g = 0.1g.

The ground motion values apply to ground motion expected for future individual earthquakes. The probabilistic ground motion calculation takes into account all possible future ground motions from all modeled earthquake magnitudes at all possible distances from the map site.

The spatial distribution of probabilistic ground motion values is shown with contours on the map, like a topography map shows different elevations, with each color representing a different range of levels of shaking.

Annual rate of exceedance

How to get the expected number of exceedances in 1 year owing to that earthquake.

Multiply the annual occurrence rate of the earthquake times the probability of exceedance of the ground motion, given that earthquake.

Expected number of exceedances in 1 year = Annual rate of exceedance

Annual rate of exceedance, given several earthquakes

Expected number of exceedances for several earthquakes. (Adding exceedances)

The expected number of exceedances for several earthquakes is calculated by merely adding the annual rate of exceedance owing to each earthquake.

Calculating a hazard curve

A hazard curve is calculated by plotting annual rate of exceedance vs ground motion

a. Perform the above calculation for 18 other ground motion levels.
b. Plot the results.
c. Make a smooth curve.

Now, for any ground motion we can find the annual rate of exceedance. Likewise, for any annual rate of exceedance we can find the corresponding ground motion.

Now one can use the hazard curve to find the corresponding ground motion. The hazard maps are just the contoured version of the corresponding ground motion plotted on a geographic grid.

COMPTENCY 16.0 UNDERSTAND GEOLOGIC MAPPING.

SKILL 16.1 Know how to find position using a topographic map

Topographic maps use fine lines drawn in ordered patterns to show the topography and elevation of the land.

Topographic Map Symbols
A series of special symbols and lines are used to display information about the shape and elevation of the landscape. There is a set of rules that determine how these symbols and lines are drawn on the map.
- **Contour Interval**: Shows the amount of elevation between contour lines.
- **Bench Mark**: Shows exact elevation. Often marked with a solid triangle.
- **Contour Line**: Connects points that have the same elevation. Contour lines are closed loops although all of the loop may not be visible on the map.
- **Index Contours**: The heaviest contour lines, each marked with an elevation.
- **Elevation**: A numerical indication of the contour line's elevation.
- **Hachure Marks**: Short lines drawn inside a closed loop that indicate a depression. The marks point down slope.
- **Gradient**: The relative spacing of the contouring lines indicates the gradient of the slope. The closer the lines, the steeper the gradient.
- **Sea Level**: Elevation is measured as either above or below sea level.

The Five General Rules of Contouring
1. All contour lines either close or extend to the edge of the map.
2. Contour lines are closed around hills, basins, or depressions.
3. Contour lines never cross, although they are sometimes very close. The closer they are, the steeper the slope.
4. Contour lines appear on both sides of an area where the slope reverses direction.
5. Contour lines form V's hat point upstream when they cross a stream, river, or valley.

The topographic map **legend** provides a great deal of information about the map, including the scale of the map, the agency that created the map, and the year it was created. In the upper right hand corner of the map, another legend provides the geographic name of the area covered, latitude and longitude information, and the **minute series**-the relative coverage of the map. The larger the minute series number, the larger the area covered by the map. If a topographic map shows open bodies of water, the elevation markings in the water are given as depth soundings. You can determine a great deal of information about the geologic processes that have shaped the landscape by carefully analyzing topographic maps.

SKILL 16.2 Know how to make a geologic map showing faults, structural data, and contacts between formations.

A **geologic map** is a special-purpose map made to show subsurface geological features. In the United States, geologic maps are usually superimposed over a topographic map (and at times over other base maps) with the addition of a color mask with letter symbols to represent the kind of geologic unit, stratigraphic contour lines, fault lines, strike and dip symbols, and various additional symbols as indicated by the map key.

The most striking features of geologic maps are its colors. Each color represents a different geologic unit. A geologic unit is a volume of a certain kind of rock of a given age range. A sandstone of one age might be colored bright orange, while a sandstone of a different age might be colored pale brown. The capital letter represents the age of the geologic unit. Geologists have divided the history of the Earth into Eons, Eras, Periods, and Epochs. The most common division of time used in letter symbols on geologic maps is the Period. Most letter symbols begin with a capital letter representing one of the four Periods: J (Jurassic), K (Cretaceous), T (Tertiary), or Q (Quaternary). Sometimes the age of a rock unit will span more than one period. In that case, both capital letters are used. For example, QT would indicate that the rock unit began to form in Tertiary time and was completed in Quaternary time. The small letters indicate either the name of the unit, if it has one, or the type of rock, if the unit has no name.

The place where two different geologic units are found next to each other is called a contact, and that is represented by different kinds of lines on the geologic map. The two main types of contacts shown on most geologic maps are depositional contacts and faults. All geologic units are formed over, under, or beside other geologic units. When different geologic units have been moved next to one another after they were formed, the contact is a fault contact, which is shown on the map by a thick line. Another kind of line shown on most geologic maps is a fold axis. In addition to being moved by faults, geologic units can also be bent and warped by the same forces into rounded wavelike shapes called folds. A line that follows the crest or trough of the fold is called the fold axis. This is marked on a geologic map with a line a little thicker than a depositional contact, but thinner than a fault.

All thicknesses of lines are also modified by being solid, dashed, or dotted. Often contacts are obscured by soil, vegetation, or human construction. Those places where the line is precisely located it is shown as solid, but where it is uncertain it is dashed. The shorter the dash, the more uncertain the location. The lines on the map may also be modified by other symbols on the line (triangles, small tic marks, arrows, etc.) which give more information about the line. For example, faults with triangles on them show that the side with the triangles has been thrust up and over the side without the triangles. All the different symbols on the lines are explained in the map key.

A geologic map shows the distribution of rocks at the earth's surface. However, bedrock is usually obscured so that only a small amount of outcrop is available for observation, study, or sampling. The geologist must then extrapolate the general distribution of rock types. Not only will the rock he can see help him, but so will changes in soil, vegetation, and landscape as well as various patterns detected from aerial photographs.

A geologic cross section map tells how the rocks are arranged underground. Each type of rock has its own symbol for shading and various structures have defined symbols. The geologic map and its cross section are used together to give a more complete picture.

SKILL 16.3 Know how to interpret geologic history and processes from a geologic map.

Displaying the Historical Record
When the Earth Scientist has collected the individual pieces of information, it is generally graphically displayed in Geologic Columns, Geologic Cross Sections, and Geologic Maps.

Geologic Column: a graphic display of a slice of the Earth from one locale. A geologic column shows the types of rocks found and may or may not show fossils and ages.

Geologic Cross Section: a graphic display that shows how the underground rock units are positioned in relation to each other. Geologic Cross Sections are based on the information contained in Geologic columns.

Geologic Map: a graphic display showing the structure of rock formations, the distribution of rock, and the ages of the rock that is just below the loose soil and vegetation covering it. Geologic maps may also be constructed to show a specific type of information, such as:

> **Paleogeographic Map**: Shows the geography of a past time.
>
> **Isopachus Map**: Shows the thickness (depths) of the underground rock units. Important economic utilization to hunt for oil, minerals, etc.
>
> **Lithofacies Map**: A combination of Paleogeographic and Isopachus maps, the Lithofacies Map shows the distribution of rock units in an area.

TEACHER CERTIFICATION STUDY GUIDE

PART II: SUBJECT MATTER SKILLS AND ABILITIES APPLICABLE TO THE CONTENT DOMAINS IN SCIENCE

DOMAIN 1. INVESTIGATION AND EXPERIMENTATION

Skill 1.1 Question Formulation

The first step in scientific inquiry is posing a question to be answered. Next, one forms a hypothesis, and then conducts an experiment to test the hypothesis. Comparison between the predicted and observed results is the next step. Conclusions are then formed based on the analysis and it is determined whether the hypothesis is correct or incorrect. If incorrect, the next step is to form a new hypothesis and the process is repeated.

Let's use the following everyday situation as an example. Through the course of making breakfast, you bring three eggs from the refrigerator over to the stove. Your hands are full and you accidentally drop an egg on the floor, which immediately shatters all over the tile floor. As you clean up the mess you wonder if you had carried the eggs in their cardboard container, would they have broken if dropped? Similarly, if dropped would they have broken on a softer surface, for example linoleum?

a. Formulate and evaluate a viable hypothesis

Once the question is formulated take an educated guess about the answer to the problem or question. For our scientist above, a plausible hypothesis might be that even if dropped, the egg would not have broken if it had been enclosed in its protective cardboard box.

b. Recognize the value and role of observation prior to question formulation

The scientist conducting our imaginary egg experiment made observations prior to the experiment. He knows that eggshells are fragile, and that their interior is liquid. He also noted that his floor was made of tile, a hard surface, and that the broken egg had not been protected. His observations, however general they may have seemed, led him to create a viable question and an educated guess (hypothesis) about what he expected. While scientists often have laboratories set up to study a specific thing, it is likely that along the way they will find an unexpected result. It is always important to be open-minded and to look at all of the information. An open-minded approach to science provides room for more questioning, and, hence, more learning.

c. Recognize the iterative nature of questioning

The question stage of scientific inquiry involves repetition. By repeating the experiment you can discover whether or not you have reproducibility. If results are reproducible, the hypothesis is valid. If the results are not reproducible, one has more questions to ask.

d. Given an experimental design, identify possible hypotheses that it may test

An experiment is proposed and performed with the sole objective of testing a hypothesis. You discover the aforementioned scientist conducting an experiment with the following characteristics. He has two rows each set up with four stations. The first row has a piece of tile as the base at each station. The second row has a piece of linoleum as the base at each station. The scientist has eight eggs and is prepared to drop one over each station. What is he testing? He is trying to answer whether or not the egg is more likely to break when dropped over one material as opposed to the other. His hypothesis might have been: The egg will be less likely to break when dropped on linoleum.

Skill 1.2 Planning a Scientific Investigation (including Experimental Design)

a. Given a hypothesis, formulate an investigation or experimental design to test that hypothesis

Suppose our junior scientist wants to look at his initial question, "if you had carried the eggs in their cardboard container, would they have broken if dropped?" A sensible hypothesis to this question would be that an egg would be less likely to break if it was dropped in its cardboard container, than if it were unprotected. Because reproducibility is important, we need to set up multiple identical stations, or use the same station for repeatedly conducting the same experiment. Either way it is key that everything is identical. If the scientist wants to study the break rate for one egg in it's container, then it needs to be just one egg dropped each time in an identical way. The investigator should systematically walk to each station and drop an egg over each station and record the results. The first four times, the egg should be dropped without enclosing it in a cardboard carton. This is the control. It is a recreation of what happened accidentally in the kitchen and one would expect the results to be the same- an egg dropped onto tile will break. The next four times, the egg should be dropped nestled within its original, store manufactured, cardboard container. One would expect that the egg would not break, or would break less often under these conditions.

b. Evaluate an experimental design for its suitability to test a given hypothesis

When designing an experiment, one needs to clearly define what one is testing. One also needs to consider the question asked. The more limited the question, the easier it is to set up an experiment to answer it. Ideally, if an egg were dropped, the egg would be safest when dropped in a protective carton over a soft surface. However, one should not measure multiple variables at once. Studying multiple variables at once makes the results difficult to analyze. How would the investigator discern which variable was responsible for the result? When evaluating experimental design, make sure to look at the number of variables, how clearly they were defined, and how accurately they were measured. Also, was the experiment applicable? Did it make sense and address the hypothesis?

c. Distinguish between variable and controlled parameters

The procedure used to obtain data is important to the outcome. Experiments consist of **controls** and **variables**. A control is the experiment run under normal conditions. The variable includes a factor that is changed. In biology, the variable may be light, temperature, pH, time, etc. The differences in tested variables may be used to make a prediction or form a hypothesis. Only one variable should be tested at a time. One would not alter both the temperature and pH of the experimental subject.

An **independent variable** is one that is changed or manipulated by the researcher. This could be the amount of light given to a plant or the temperature at which bacteria is grown. The **dependent variable** is that which is influenced by the independent variable.

Skill 1.3   Observation and Data Collection

a. Identify changes in natural phenomena over time without manipulating the phenomena (e.g. a tree limb, a grove of trees, a stream, a hill slope).

Scientists identify changes in natural phenomena over time using basic tools of measurement and observation. Scientists measure growth of plants by measuring plant dimensions at different time intervals, changes in plant and animal populations by counting, and changes in environmental conditions by observation. The following are four examples of natural phenomena, and the observation techniques used to measure change in each case.

To identify change in a tree limb, we measure the dimensions (length, circumference) of the limb at different time intervals. In addition, we can study the types and amount of organisms growing on the limb by observing a small sample and applying the observations to make estimations about the entire limb. Finally, we can watch for the presence of disease or bacterial infection by observing the color and consistency of the limb and any changes over time.

To identify change in a grove of trees, we employ similar techniques as used in the observation of a tree limb. First, we measure the size of the trees at different time intervals. If the grove contains many trees, we may measure only a representative sample of trees and apply the results to make conjectures about the grove population. Finally, we closely monitor the trees for changes that may indicate disease or infection.

To identify change in a stream, we measure and observe characteristics of both the stream itself and the organisms living in it. First, we measure the width and depth of the stream at different time intervals to monitor erosion. Second, we observe the water level at different time intervals to monitor the effect of weather patterns. Finally, using sampling techniques, we observe and measure the types and number of organisms present in the stream and how these characteristics change over time.

To identify change on a hill slope, we measure the angle and dimensions of the slope at different time intervals to monitor the effects of erosion by wind and rain. In addition, we use sampling techniques to make generalizations about the organisms living on the slope. Finally, we can monitor how the types and amounts of vegetation on the slope change in relation to the change in the angle of the slope (i.e. determine which types of plants have the ability to grow in certain conditions).

b. **Analyze the locations, sequences, and time intervals that are characteristic of natural phenomena (e.g. locations of planets over time, succession of species in an ecosystem).**

One of the main goals of science is the study and explanation of natural phenomena. When studying natural phenomena, scientists describe the characteristic locations, sequences, and time intervals. Examples of natural phenomena studied by scientists include the locations of planets over time and the succession of species in an ecosystem.

The eight planets of the solar system (Pluto was formerly included as a planet but has been removed as of Summer 2006) orbit the sun in a specific sequence. The time it takes to complete an orbit of the Sun is different for each planet. In addition, we can determine the location of each planet in relation to the Sun and to each other using mathematical models and charts.

Mercury orbits closest to the sun, followed by Venus, Earth, Mars, Jupiter, Saturn, Uranus, and Neptune. Neptune is farthest from the Sun for 20 of every 248 years. Planets will never collide because one is always higher than the other, even when their orbits do intersect.

The amount of time a planet takes to complete one orbit of the Sun increases as the distance from the Sun increases. This value, called the sidereal period, ranges from 0.241 years for Mercury to 248.1 years. The synodic period measures the amount of time it takes for a planet to return to the same point in the sky as observed from Earth. Mercury has the shortest synodic period of 116 days while Mars has the longest of 780 days. The synodic periods of Jupiter, Saturn, Uranus, and Neptune are similar, slightly less than 400 days for each.

Succession of species is the change in the type and number of plants, animals, and microorganisms that occurs periodically in all ecosystems. The two types of succession are primary and secondary. Primary succession describes the creation and subsequent development of a new, unoccupied habitat (e.g. a lava flow). Secondary succession describes the disruption of an existing community (e.g. fire, human tampering, flood) and the response of the community to the disruption. Succession is usually a very long process. New communities often take hundreds or thousands of years to reach a fully developed state (climax community). And, while succession in climax communities is minimal, environmental disruption can easily restart the succession process.

In general, simple organisms (e.g. bacteria, small plants) dominate new communities and prepare the environment for the development of larger, more complex species. For example, the dominant vegetation of an empty field will progress sequentially from grasses to small shrubs to soft wood trees to hard wood trees. We can observe and measure succession in two ways. First, we can measure the changes in a single community over time. Second, we can observe and compare similar communities at different stages of development. We are limited in the amount of data we can gather using the first method because of the slow nature of the succession process. The techniques used to observe succession include fossil observation, geological study, and environmental sampling.

c. Select and use appropriate tools and technology (e.g. computer-linked probes, spreadsheets, graphing calculators) to perform tests, collect data, analyze relationships, and display data.

Scientists use a variety of tools and technologies to perform tests, collect and display data, and analyze relationships. Examples of commonly used tools include computer-linked probes, spreadsheets, and graphing calculators.

Scientists use computer-linked probes to measure various environmental factors including temperature, dissolved oxygen, pH, ionic concentration, and pressure. The advantage of computer-linked probes, as compared to more traditional observational tools, is that the probes automatically gather data and present it in an accessible format. This property of computer-linked probes eliminates the need for constant human observation and manipulation.

Scientists use spreadsheets to organize, analyze, and display data. For example, conservation ecologists use spreadsheets to model population growth and development, apply sampling techniques, and create statistical distributions to analyze relationships. Spreadsheet use simplifies data collection and manipulation and allows the presentation of data in a logical and understandable format.

Graphing calculators are another technology with many applications to biology. For example, scientists use algebraic functions to analyze growth, development and other natural processes. Graphing calculators can manipulate algebraic data and create graphs for analysis and observation. In addition, scientists use the matrix function of graphing calculators to model problems in genetics. The use of graphing calculators simplifies the creation of graphical displays including histograms, scatter plots, and line graphs. Scientists can also transfer data and displays to computers for further analysis. Finally, scientists connect computer-linked probes, used to collect data, to graphing calculators to ease the collection, transmission, and analysis of data.

d. Evaluate the precision, accuracy, and reproducibility of data

Accuracy is the degree of conformity of a measured, calculated quantity to its actual (true) value. Precision also called reproducibility or repeatability and is the degree to which further measurements or calculations will show the same or similar results. Accuracy is the degree of veracity while precision is the degree of reproducibility. The best analogy to explain accuracy and precision is the target comparison. Repeated measurements are compared to arrows that are fired at a target. Accuracy describes the closeness of arrows to the bull's eye at the target center. Arrows that strike closer to the bull's eye are considered more accurate.

e. Identify and analyze possible reasons for inconsistent results, such as sources of error or uncontrolled conditions

Reproducibility is highly important when considering science. If results are not reproducible, they are usually not given much credit, regardless of the hypothesis. For this reason, we pay close attention to minimizing sources of error. Examples of common sources of error might be contamination or an improperly mixed buffer. In addition, one should remember that scientists are humans, and human error is always a possibility.

All experimental uncertainty is due to either random errors or systematic errors. Random errors are statistical fluctuations in the measured data due to the precision limitations of the measurement device. Random errors usually result from the experimenter's inability to take the same measurement in exactly the same way to get exactly the same number. Systematic errors, by contrast, are reproducible inaccuracies that are consistently in the same direction. Systematic errors are often due to a problem, which persists throughout the entire experiment. Systematic and random errors refer to problems associated with making measurements. Mistakes made in the calculations or in reading the instrument are not considered in error analysis.

f. Identify and communicate sources of unavoidable experimental error.

Unavoidable experimental error is the random error inherent in scientific experiments regardless of the methods used. One source of unavoidable error is measurement and the use of measurement devices. Using measurement devices is an imprecise process because it is often impossible to accurately read measurements. For example, when using a ruler to measure the length of an object, if the length falls between markings on the ruler, we must estimate the true value. Another source of unavoidable error is the randomness of population sampling and the behavior of any random variable. For example, when sampling a population we cannot guarantee that our sample is completely representative of the larger population. In addition, because we cannot constantly monitor the behavior of a random variable, any observations necessarily contain some level of unavoidable error.

g. Recognize the issues of statistical variability and explain the need for controlled tests.

Statistical variability is the deviation of an individual in a population from the mean of the population. Variability is inherent in biology because living things are innately unique. For example, the individual weights of humans vary greatly from the mean weight of the population. Thus, when conducting experiments involving the study of living things, we must control for innate variability. Control groups are identical to the experimental group in every way with the exception of the variable being studied. Comparing the experimental group to the control group allows us to determine the effects of the manipulated variable in relation to statistical variability.

h. Know and evaluate the safety issues when designing an experiment and implement appropriate solutions to safety problems

All science labs should contain the following items of **safety equipment**. Those marked with an asterisk are requirements by state laws.

* fire blanket which is visible and accessible
*Ground Fault Circuit Interrupters (GCFI) within two feet of water supplies
*signs designating room exits
*emergency shower providing a continuous flow of water
*emergency eye wash station which can be activated by the foot or forearm
*eye protection for every student and a means of sanitizing equipment
*emergency exhaust fans providing ventilation to the outside of the building
*master cut-off switches for gas, electric and compressed air. Switches must have
  permanently attached handles. Cut-off switches must be clearly labeled.
*an ABC fire extinguisher
*storage cabinets for flammable materials
-chemical spill control kit
-fume hood with a motor which is spark proof
-protective laboratory aprons made of flame retardant material
-signs which will alert potential hazardous conditions
-containers for broken glassware, flammables, corrosives, and waste.
- containers should be labeled.

Students should wear safety goggles when performing dissections, heating, or while using acids and bases. Hair should always be tied back and objects should never be placed in the mouth. Food should not be consumed while in the laboratory. Hands should always be washed before and after laboratory experiments. In case of an accident, eye washes and showers should be used for eye contamination or a chemical spill that covers the student's body. Small chemical spills should only be contained and cleaned by the teacher. Kitty litter or a chemical spill kit should be used to clean spill. For large spills, the school administration and the local fire department should be notified. Biological spills should also be handled only by the teacher. Contamination with biological waste can be cleaned by using bleach when appropriate.

Accidents and injuries should always be reported to the school administration and local health facilities. The severity of the accident or injury will determine the course of action to pursue.

It is the responsibility of the teacher to provide a safe environment for their students. Proper supervision greatly reduces the risk of injury and a teacher should never leave a class for any reason without providing alternate supervision. After an accident, two factors are considered; **foreseeability** and **negligence**. Foreseeability is the anticipation that an event may occur under certain circumstances. Negligence is the failure to exercise ordinary or reasonable care. Safety procedures should be a part of the science curriculum and a well managed classroom is important to avoid potential lawsuits.

i. **Appropriately employ a variety of print and electronic resources (e.g. the World Wide Web) to collect information and evidence as part of a research project.**

Scientists use print and electronic resources to collect information and evidence. Gathering information from scientific literature is a necessary element in successful research project design. Scientific journals, a major source of scientific information, provide starting points for experimental design and points of comparison in the interpretation of experimental results. Examples of important scientific journals are *Science*, *Nature*, and *Cell*. Scientists use the World Wide Web to search and access scientific journal articles through databases such as PubMed, JSTOR, and Google Scholar. In addition, the World Wide Web is a rich source of basic background information useful in the design and implementation of research projects. Examples of relevant online resources include scientific encyclopedias, general science websites, and research laboratory homepages.

j. **Assess the accuracy validity and reliability of information gathered from a variety of sources**

Because people often attempt to use scientific evidence in support of political or personal agendas, the ability to evaluate the credibility of scientific claims is a necessary skill in today's society. In evaluating scientific claims made in the media, public debates, and advertising, one should follow several guidelines.

First, scientific, peer-reviewed journals are the most accepted source for information on scientific experiments and studies. One should carefully scrutinize any claim that does not reference peer-reviewed literature.

Second, the media and those with an agenda to advance (advertisers, debaters, etc.) often overemphasize the certainty and importance of experimental results. One should question any scientific claim that sounds fantastical or overly certain.

Finally, knowledge of experimental design and the scientific method is important in evaluating the credibility of studies. For example, one should look for the inclusion of control groups and the presence of data to support the given conclusions.

TEACHER CERTIFICATION STUDY GUIDE

Skill 1.4   Data Analysis/Graphing

a. Construct appropriate graphs from data and develop qualitative and quantitative statements about relationships between variables

Graphing is an important skill to visually display collected data for analysis. The two types of graphs most commonly used are the **line graph** and the **bar graph** (histogram). Line graphs are set up to show two variables represented by one point on the graph. The X-axis is the horizontal axis and represents the dependent variable. Dependent variables are those that would be present independently of the experiment. A common example of a dependent variable is time. Time proceeds regardless of anything else going on. The Y-axis is the vertical axis and represents the independent variable. Independent variables are manipulated by the experiment, such as the amount of light, or the height of a plant. Graphs should be calibrated at equal intervals. If one space represents one day, the next space may not represent ten days. A "best fit" line is drawn to join the points and may not include all the points in the data. Axes must always be labeled. A good title will describe both the dependent and the independent variable. Bar graphs are set up similarly in regards to axes, but points are not plotted. Instead, the dependent variable is set up as a bar where the X-axis intersects with the Y-axis. Each bar is a separate item of data and is not joined by a continuous line.

When drawing conclusions from graphs, one can make quantitative or qualitative statements. Quantitative is derived from quantity (numerical, precise) and qualitative (impressive) is derived from quality.   For example, stating that the median is 12 would be a quantitative assessment.

The type of graphic representation used to display observations depends on the data that is collected. **Line graphs** are used to compare different sets of related data or to predict data that has not yet be measured. An example of a line graph would be comparing the rate of activity of different enzymes at varying temperatures. A **bar graph** or **histogram** is used to compare different items and make comparisons based on this data. An example of a bar graph would be comparing the ages of children in a classroom. A **pie chart** is useful when organizing data as part of a whole. A good use for a pie chart would be displaying the percent of time students spend on various after school activities.

b. Recognize the slope of the linear graph as the constant in the relationship y=kx and apply this principle in interpreting graphs constructed from data

Analyzing graphs is a useful method for determining the mathematical relationship between the dependent and independent variables of an experiment. The usefulness of the method lies in the fact that the variables represents on the axes of a straight-line graph are represented by the expression, $y = m*x + b$, where m=the slope of the line, b=the y intercept of the line. This equation works only if the data fit a straight-line graph. Thus, once the data set has been collected, and modified, and plotted to achieve a straight-line graph, the mathematical equation can be derived.

c. Apply simple mathematical relationships to determine a missing quantity in an algebraic expression, given the two remaining terms (e.g., speed = distance/time, density = mass/volume, force = pressure x area, volume = area x height)

Science and mathematics are related. Science data is strongest when accurate, and is therefore described in terms of units. To acquire proper units, one must apply math skills. Some common examples include speed, density, force, and volume. Let us look at density –

$D = m/v$
Where
D = density g/cm
m = mass in grams
v = volume in cm

One would substitute known quantities for the alphabetical symbols. It is absolutely important to write the appropriate units e.g., g (gram), cm (centimeter) etc. This is fundamental algebra.

The second example is the formula for calculating momentum of an object.

M = mass (kg) times velocity (meters/second)
$M = mv$
The units of momentum are kg (m/s)

d. Determine whether a relationship on a given graph is linear or non-linear and determine the appropriateness of extrapolating the data.

The individual data points on the graph of a linear relationship cluster around a line of best fit. In other words, a relationship is linear if we can sketch a straight line that roughly fits the data points. Consider the following examples of linear and non-linear relationships.

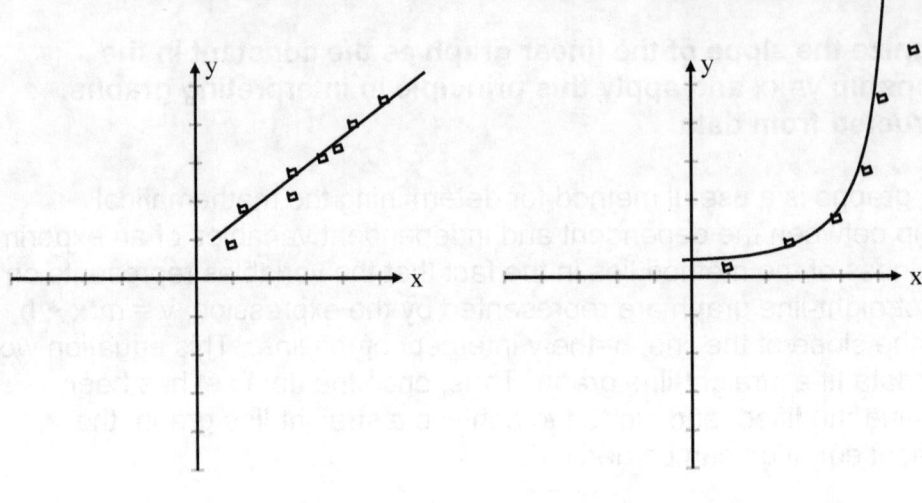

Linear Relationship   Non-Linear Relationship

Note that the non-linear relationship, an exponential relationship in this case, appears linear in parts of the curve. In addition, contrast the preceding graphs to the graph of a data set that shows no relationship between variables.

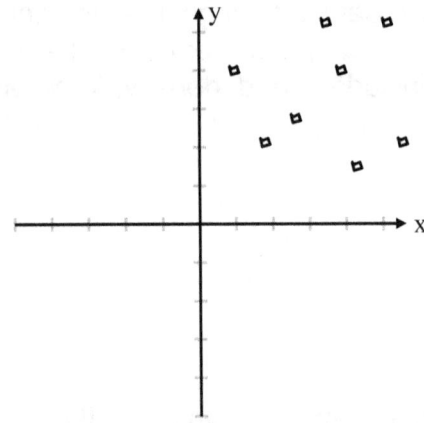

Extrapolation is the process of estimating data points outside a known set of data points. When extrapolating data of a linear relationship, we extend the line of best fit beyond the known values. The extension of the line represents the estimated data points. Extrapolating data is only appropriate if we are relatively certain that the relationship is indeed linear. For example, the death rate of an emerging disease may increase rapidly at first and level off as time goes on. Thus, extrapolating the death rate as if it were linear would yield inappropriately high values at later times. Similarly, extrapolating certain data in a strictly linear fashion, with no restrictions, may yield obviously inappropriate results. For instance, if the number of plant species in a forest were decreasing with time in a linear fashion, extrapolating the data set to infinity would eventually yield a negative number of species, which is clearly unreasonable.

e. Solve scientific problems by using quadratic equations and simple trigonometric, exponential, and logarithmic functions.

Scientists use mathematical tools and equations to model and solve scientific problems. Solving scientific problems often involves the use of quadratic, trigonometric, exponential, and logarithmic functions.

Quadratic equations take the standard form $ax^2 + bx + c = 0$. The most appropriate method of solving quadratic equations in scientific problems is the use of the quadratic formula. The quadratic formula produces the solutions of a standard form quadratic equation.

$$x = \frac{-b \pm \sqrt{b^2 - 4ac}}{2a} \quad \text{\{Quadratic Formula\}}$$

One common application of quadratic equations is the description of biochemical reaction equilibriums. Consider the following problem.

Example 1

80.0 g of ethanoic acid (MW = 60g) reacts with 85.0 g of ethanol (MW = 46g) until equilibrium. The equilibrium constant is 4.00. Determine the amounts of ethyl acetate and water produced at equilibrium.

$$CH_3COOH + CH_3CH_2OH = CH_3CO_2C_2H_5 + H_2O$$

The equilibrium constant, K, describes equilibrium of the reaction, relating the concentrations of products to reactants.

$$K = \frac{[CH_3CO_2C_2H_5][H_2O]}{[CH_3CO_2H][CH_3CH_2OH]} = 4.00$$

The equilibrium values of reactants and products are listed in the following table.

	CH_3COOH	CH_3CH_2OH	$CH_3CO_2C_2H_5$	H_2O
Initial	80/60 = 1.33 mol	85/46 = 1.85 mol	0	0
Equilibrium	1.33 − x	1.85 − x	x	x

Thus, $K = \dfrac{[x][x]}{[1.33-x][1.85-x]} = \dfrac{x^2}{2.46 - 3.18x + x^2} = 4.00$.

Rearrange the equation to produce a standard form quadratic equation.

$$\frac{x^2}{2.46-3.18x+x^2} = 4.00$$

$$x^2 = 4.00(2.46-3.18x+x^2) = 9.84-12.72x+4x^2$$

$$0 = 3x^2 - 12.72x + 9.84$$

Use the quadratic formula to solve for x.

$$x = \frac{-(-12.72) \pm \sqrt{(-12.72)^2 - 4(3)(9.84)}}{2(3)} = 3.22 \text{ or } 1.02$$

3.22 is not an appropriate answer, because we started with only 3.18 moles of reactants. Thus, the amount of each product produced at equilibrium is 1.02 moles.

Scientists use trigonometric functions to define angles and lengths. For example, field scientists can use trigonometric functions to estimate distances and directions. The basic trigonometric functions are sine, cosine, and tangent. Consider the following triangle describing these relationships.

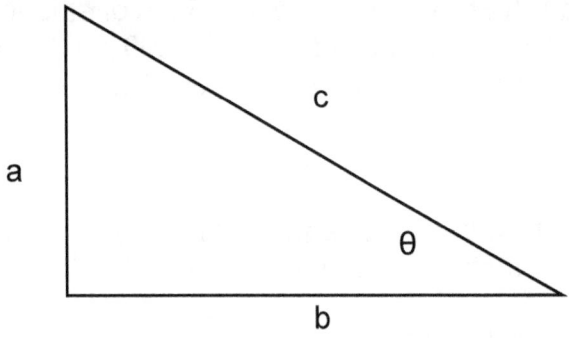

$$\sin \theta = \frac{a}{c}, \quad \cos \theta = \frac{b}{c}, \quad \tan \theta = \frac{a}{b}$$

Exponential functions are useful in modeling many scientific phenomena. For example, scientists use exponential functions to describe bacterial growth and radioactive decay. The general form of exponential equations is $f(x) = Ca^x$ (C is a constant). Consider the following problem involving bacterial growth.

Example 2

Determine the number of bacteria present in a culture inoculated with a single bacterium after 24 hours if the bacterial population doubles every 2 hours. Use $N(t) = N_0 e^{kt}$ as a model of bacterial growth where N(t) is the size of the population at time t, N_0 is the initial population size, and k is the growth constant.

We must first determine the growth constant, k. At t = 2, the size of the population doubles from 1 to 2. Thus, we substitute and solve for k.

$$2 = 1(e^{2k})$$

$\ln 2 = \ln e^{2k}$ Take the natural log of each side.

$\ln 2 = 2k(\ln e) = 2k$ $\ln e = 1$

$k = \dfrac{\ln 2}{2}$ Solve for k.

The population size at t = 24 is

$$N(24) = e^{(\frac{\ln 2}{2})24} = e^{12\ln 2} = 4096.$$

Finally, logarithmic functions have many applications to science and biology. One simple example of a logarithmic application is the pH scale. Scientists define pH as follows.

$pH = -\log_{10} [H_+]$, where $[H_+]$ is the concentration of hydrogen ions

Thus, we can determine the pH of a solution with a $[H_+]$ value of 0.0005 mol/L by using the logarithmic formula.

$pH = -\log_{10} [0.0005] = 3.3$

Skill 1.5   Drawing Conclusions and Communicating Explanations

The state of California needs to ensure that its licensed teachers are capable of the list below. These items are not items that can be explained in essay format; rather they are an accumulation of your years of learning. You will be able to find correlations with these items in other areas of this manual.

a. Draw appropriate and logical conclusions from data
b. Communicate the logical connection among hypotheses, science concepts, tests conducted, data collected, and conclusions drawn from the scientific evidence
c. Communicate the steps and results of an investigation in written reports and oral presentations
d. Recognize whether evidence is consistent with a proposed explanation
e. Construct appropriate visual representations of scientific phenomenon and processes (e.g., motion of Earth's plates, cell structure)
f. Read topographic and geologic maps for evidence provided on the maps and construct and interpret a simple scale map

DOMAIN 2. NATURE OF SCIENCE

Skill 2.1 Scientific Inquiry

a. Distinguish among the terms hypothesis, theory, and prediction as used in scientific investigations

Science may be defined as a body of knowledge that is systematically derived from study, observations, and experimentation. Its goal is to identify and establish principles and theories that may be applied to solve problems. Pseudoscience, on the other hand, is a belief that is not warranted. There is no scientific methodology or application. Some of the more classic examples of pseudoscience include witchcraft, alien encounters or any topic that is explained by hearsay.

Scientific theory and experimentation must be repeatable. It is also possible to be disproved and is capable of change. Science depends on communication, agreement, and disagreement among scientists. It is composed of theories, laws, and hypotheses.

theory - the formation of principles or relationships which have been verified and accepted.

law - an explanation of events that occur with uniformity under the same conditions (laws of nature, law of gravitation).

hypothesis - an unproved theory or educated guess followed by research to best explain a phenomena. A theory is a proven hypothesis.

Science is limited by the available technology. An example of this would be the relationship of the discovery of the cell and the invention of the microscope. As our technology improves, more hypotheses will become theories and possibly laws. Science is also limited by the data that is able to be collected. Data may be interpreted differently on different occasions. Science limitations cause explanations to be changeable as new technologies emerge.

The first step in scientific inquiry is posing a question to be answered. Next, a hypothesis is formed to provide a plausible explanation. An experiment is then proposed and performed to test this hypothesis. A comparison between the predicted and observed results is the next step. Conclusions are then formed and it is determined whether the hypothesis is correct or incorrect. If incorrect, the next step is to form a new hypothesis and the process is repeated.

b. **Evaluate the usefulness, limitations, and interdisciplinary and cumulative nature of scientific evidence as it relates to the development of models and theories as representations of reality**

All evidence can be manipulated by the presenter for their own purposes. This is why we stress that one must carefully evaluate resources. For instance, when reading a scientific article: Is it published in a well known journal, does it use controls, does it make sense, is the experiment clearly explained, are the results reproducible? One must also recognize the limitations of research. An experiment is more clearly analyzed if it only has one variable. Would the research still be true if another variable, for example, heat, time, or substrate were changed? One must consider the conditions under which the research was conducted. Were the most advanced technological machines used, or would there be a more applicable way to study the issue? For example, no one realized there was more to know about microscopic life until microscopy became more advanced. We now use scanning electron microscopes (SEM's), making light microscopes somewhat obsolete, and opening our eyes to a whole new level of thoroughness. As technology changes, so too does our knowledge and our awareness of reality. Galileo was a major scientist of his time (often referred to as the father of science) and used mathematics to properly describe scientific events. For all of his great efforts, though, as our machines have grown in power, we have had to rethink some of his theories. His improvements on the telescope enabled him to locate and accurately name many planets, stars, and systems. He was unable, however, to correctly ascertain the orbits of planets and the genesis of tides. Sir Isaac Newton expounded upon previous works, including Galileo's, when creating his laws of physics. Thus, tides were finally explained accurately, through an accumulation of knowledge.

c. **Recognize that when observations do not agree with an accepted scientific theory, either the observations are mistaken or fraudulent, or the accepted theory is erroneous or incorrect**

Sir Isaac Newton must have sensed that Galileo's tide theory didn't make sense- it didn't hold up to his observations. He had the opportunity, like present day scientists, to review his observations for error, or find a better explanation. One must note, though, that better in this case must be scientifically accurate in order to be impressive to peers specializing in science.

d. **Understand that reproducibility of data is critical to the scientific endeavor**

In order to have your theory accepted, it must be accurate and clearly derived. This means that another scientist could recreate your experiment from your notes, find similar data, and draw the same conclusions. In this way the validity of science is substantiated.

e. **Recognize that science is a self-correcting process that eventually identifies misconceptions and experimental biases**

The scientific process encourages periodic reassessment. The conclusion step allows one to examine the hypothesis as it relates to their experimental data. At this point, one can find positive correlations or discord. When results are unexpected, one should revisit all possible sources of error. If an error is not found to explain the results, one can reconsider the hypothesis and also think about other possibilities. This is why experimentation often results in further experimentation.

f. **Recognize that an inquiring mind is at the heart of the scientific method and that doing science involves thinking critically about the evidence presented, the usefulness of models, and the limitations of theories**

Science is not merely about creating; it is also about assessment and solutions. Science can be thought of as a loop. One questions something, and creates an experiment to study it. One can learn from this evidence, and then ask more questions. In depth learning involves looking at the experimental data from all angles and continuing to seek knowledge. Learning in depth does not occur by looking at something superficially or by taking someone else's data as 'proof.' Go one step further: analyze the evidence as if you were searching for a problem- maybe there won't be one, but you will be more likely to find it if there is!

g. **Recognize that theories are judged by how well they explain observations and predict results and that when they represent new ideas that are counter to mainstream ideas they often encounter vigorous criticism**

If a theory explains a phenomenon well, it is worth considering, even if it turns out to be incorrect later on. The problem with this is two fold. First, a person can use a theory to push their own beliefs. This is the case with people seeing what they want to see, and then forming theories based around their opinions. An example would be if a scientist expected certain results, and then found ways to skew the results to match his theory. A theory based upon opinions will soon be seen as transparent and will be dismissed because it has no pertinent data to support it. Even if a theory is developed well, it still may not be readily accepted. A new theory is almost always difficult to introduce to an established community. To have a theory hold up to scrutiny, the author must have accurate data. Second, the author must continue to publicize the information. Just because a theory is not commonplace, does not mean it is incorrect. Novel ideas often become cornerstones in understanding, but it doesn't happen overnight. If the experiment has reproducible results and strong mathematics, people will eventually be swayed.

h. Recognize that when observations, data, or experimental results do not agree, the unexpected results are not necessarily mistakes; to discard the unusual in order to reach the expected is to guarantee that nothing but what is expected will ever be seen

Often, results other than what were expected are from an error. However, this is not always the case. Consider a scientist who has double checked his work multiple times and can find no errors. He can not explain what has happened except to assume that his theory was wrong. Maybe there is a fundamental scientific phenomenon that has yet to be explained and he couldn't possibly have known. Discoveries can occur in this way. If the scientist were to give up, he and society would lose the opportunity to learn something new. If the scientist opens his mind to the discovery, there are limitless possibilities for learning.

i. Know why curiosity, honesty, openness, and skepticism are so highly regarded in science and how they are incorporated into the way science is carried out

Curiosity fuels research. It prompts the questions that turn into scientific inquiry. Honesty is paramount to the scientific way. To put the research out there, and be true in your report of the findings, is to help mankind and cooperate in scientific endeavors. While antonyms, openness and skepticism are both necessary in the field of research. One should be humble. One should be open to others' ideas, and open to their own unexpected findings, but be critical in evaluation of the work as it was conducted. It is key to incorporate all of these traits and to conduct yourself in a respectable manner.

Notice: For those of you using the state of California topical guide, please note that they omit letters f and g. XAMonline has taken that fact into account and properly sequenced the letters.

Skill 2.2   Scientific Ethics

To understand scientific ethics, we need to have a clear understanding of ethics. Ethics is defined as a system of public, general rules for guiding human conduct (Gert, 1988). The rules are general because they are supposed to all people at all times and they are public because they are not secret codes or practices.

Philosophers have given a number of moral theories to justify moral rules, which range from utilitarianism (a theory of ethics that prescribes the quantitative maximization of good consequences for a population. It is a form of consequentialism. This theory was proposed by Mozi, a Chinese philosopher who lived during BC 471-381), Kantianism (a theory proposed by Immanuel Kant, a German philosopher who lived during 1724-1804, which ascribes intrinsic value to rational beings and is the philosophical foundation of contemporary human rights) to social contract theory (a view of the ancient Greeks which states that the person's moral and or political obligations are dependent upon a contract or agreement between them to form society).

The following are some of the guiding principles of scientific ethics:

1. Scientific Honesty: not to fraud, fabricate or misinterpret data for personal gain
2. Caution: to avoid errors and sloppiness in all scientific experimentation
3. Credit: give credit where credit is due and not to copy
4. Responsibility: only to report reliable information to public and not to mislead in the name of science
5. Freedom: freedom to criticize old ideas, question new research and freedom to research

a. Understand that honesty is at the core of scientific ethics; first and foremost is the honest and accurate reporting of procedures used and data collected.

Scientists are expected to show good conduct in their scientific pursuits. Conduct here refers to all aspects of scientific activity including experimentation, testing, education, data evaluation, data analysis, data storing, peer review, government funding, the staff, etc.

b. Know that all scientists are obligated to evaluate the safety of an investigation and ensure the safety of those performing the experiment

As a teacher, the safety of your classroom is your responsibility. One should make every effort to ensure students' safety. You will need to be aware of all potential safety concerns. Advance preparation will prepare you to take the necessary precautions related to the specific experiment. You should use the applicable MSDS and check pertinent regulations (at your place of employment as well as on the state/national levels). It will be necessary to take foreseeability and negligence into consideration. It is the responsibility of the scientist to make sure that all organisms associated with the project are kept safe. This refers to both people and animals.

c. Know the procedures for respectful treatment of all living organisms in experimentation and other investigations

No dissections may be performed on living mammalian vertebrates or birds. Lower order life and invertebrates may be used. Biological experiments may be done with all animals except mammalian vertebrates or birds. No physiological harm may result to the animal. All animals housed and cared for in the school must be handled in a safe and humane manner. Animals are not to remain on school premises during extended vacations unless adequate care is provided. Any instructor who intentionally refuses to comply with the laws may be suspended or dismissed.

Pathogenic organisms must never be used for experimentation. Students should adhere to the following rules at all times when working with microorganisms to avoid accidental contamination:

1. Treat all microorganisms as if they were pathogenic.
2. Maintain sterile conditions at all times

Dissection and alternatives to dissection

Animals that were not obtained from recognized sources should not be used. Decaying animals or those of unknown origin may harbor pathogens and/or parasites. Specimens should be rinsed before handling. Latex gloves are desirable. If not available, students with sores or scratches should be excused from the activity. Formaldehyde is likely carcinogenic and should be avoided or disposed of according to district regulations. Students objecting to dissections for moral reasons should be given an alternative assignment. Interactive dissections are available online or from software companies for those students who object to performing dissections. There should be no penalty for those students who refuse to physically perform a dissection.

Skill 2.3 Historical Perspectives

a. Discuss the cumulative nature of scientific evidence as it relates to the development of models and theories

Science is an ongoing process. There was a time when microscopes, telescopes, calculators, and computers did not exist. Their current availability has led to many discoveries. We have had the opportunity to investigate why people become sick, and the mechanisms responsible for their illnesses. We have also broadened our knowledge of physical science- the laws that govern the universe. With each new breakthrough we either build upon current knowledge, or if the new piece doesn't work with previous thoughts, we reevaluate the validity of all of the information, past and present. For this reason, models and theories are continuously evolving.

b. Recognize that as knowledge in science evolves, when observations do not support an accepted scientific theory, the observations are reconsidered to determine if they are mistaken or fraudulent, or if the accepted theory is erroneous or incomplete (e.g., an erroneous theory is the Piltdown Man fossil; an incomplete theory is Newton's laws of gravity)

When one realizes that their results do not match those previously established, the new results must be reconsidered. At this point, one of four possibilities exist. One should look closely at the new results. The first place for disagreement is the new observations- they may be mistaken. Was there an error in data collection or analysis? Repeating the experiment may yield results that more closely agree with the previous theory. If the results of the follow up experiment are the same, an observer may wonder if the new data is fraudulent (second possibility). Take for example the scientist who fabricates data, but repeatedly insists on its integrity, even though it contradicts previous studies (remember that having a current study contradict a previous one would be acceptable, providing the results were true and reproducible). Another possibility would be a problem with the previously accepted theory. An erroneous theory is one that was created with misinformation. An example of an erroneous theory would be the Piltdown Man fossil. The Piltdown Man fossil consisted of fragments of a skull and jaw bone collected in the early 1900's from a gravel pit at Piltdown, a village in England. The claim was asserted that this discovery was the fossilized remains of an unknown early form of man. In 1953 it was exposed as a forgery, and properly evaluated as the lower jaw bone of an ape combined with the skull of a fully developed, modern man. There is still some debate as to who created the forgery, but it provided quite a stir in the scientific community. The problem with an erroneous theory is that it can be believable, and then future assumptions may be based on its inaccuracy. When theories become entrenched this way it is difficult sometimes to go back and locate the error. This can be seen when studying phylogenies. If Piltdown Man was assumed to come from ancestors, and to have generations below him, the accusation of his being fraudulent sheds new light on the phylogenic tree as it was proposed. A final source for dispute would be that the original theory was incomplete, such as was true with Newton's laws of gravity. Galileo had created an erroneous theory to describe the motion of planets. It was discredited when Sir Isaac Newton established his famous laws of gravity. Newton's concept of gravity held until the beginning of the 20th century, when Einstein proposed his general theory of relativity. The key to Einstein's version is that inertia occurs when objects are in free-fall instead of when they are at rest. The theory of general relativity has been well accepted because of how its predictions have been repeatedly confirmed.

c. **Recognize and provide specific examples that scientific advances sometimes result in profound paradigm shifts in scientific theories**

A paradigm shift is a change in the underlying assumptions that define a particular scientific theory. Scientific advances, such as increased technology allowing different or more reliable data collection, sometimes result in paradigm shifts in scientific theories.

One classic example of a scientific paradigm shift is the transition from a geocentric (Earth-centered) to heliocentric (Sun-centered) model of the universe. Invention and development of the telescope allowed for greater observation of the planets and the Sun. The theory that the Sun is the center of the universe around which the planets, including the Earth, rotate gained acceptance largely because of the advances in observational technology.

Another example of a paradigm shift is the acceptance of plate tectonics as the explanation for large-scale movements in the Earth's crust. Advances in seismic imaging and observation techniques allowed for the collection of sufficient data to establish plate tectonics as a legitimate geological theory.

d. **Discuss the need for clear and understandable communication of scientific endeavors so that they may be reproduced and why reproduction of these endeavors is important**

Clear and understandable communication is essential for continuity and progress in science. When scientists complete scientific endeavors, such as research experiments, it is important that they carefully record their methods and results. Such precise communication and record keeping allows other scientists to reproduce the experiments in the future.

Reproduction of scientific endeavors is important because it simplifies the verification process. Because scientific experiments are subject to many sources of error, verification of results is essential. Scientists must verify results from scientific endeavors in order to justify the use of the acquired data in developing theories and future experiments.

In addition, clear communication of scientific endeavors allows scientists to learn from the work of others. Such sharing of information speeds the process of scientific research and development.

DOMAIN 3. SCIENCE AND SOCIETY

Skill 3.1 Science Literacy

a. Recognize that science attempts to make sense of how the natural and the designed world function

Human beings reside at the top of the food web for many reasons including physical dexterity and size, but largely because of brain power. We are thinkers, designed to be curious (as are our friends, the primates). Science is our attempt to understand the world around us, and to live within it. Science is not always accurate, and often theories are inadequate, or believed to be true only to be disproven later. Please remember that science is a man made endeavor, and you and your students should treat it as such.

b. Demonstrate the ability to apply critical and independent thinking to weigh alternative explanations of events

In section 1.3j we demonstrated the importance of assessing the validity of information. One should consider the suggestions given in 1.3j when weighing evidence. Additional information on this subject may be found in Scientific Inquiry: Section 2.1 a-k.

c. Apply evidence, numbers, patterns, and logical arguments to solve problems

Two of the most important aspects of science are data and honesty. In the scientific realm, numbers are stronger than words, so be sure to back up your comments with accurate data and examples. By using the scientific method, you will be more likely to catch mistakes, correct biases, and obtain accurate results. When assessing experimental data utilize the proper tools and mathematical concepts discussed in this guide. For an in depth review of the scientific method please visit Part II: Section 1.

d. Understand that, although much has been learned about the objects, events and phenomena in nature, there are many unanswered questions, i.e., science is a work in progress

The combination of science, mathematics and technology forms the scientific endeavor and makes science a success. It is impossible to study science on its own without the support of other disciplines like mathematics, technology, geology, physic and other disciplines as well. Science is an ongoing process involving multiple fields and individuals. Technology also plays a role in scientific discoveries- we are limited by technology. We are constantly creating new devices for experimentation, and with each one comes new revelations. As such, science is constantly developing. The nature of science mainly consists of three important things:

The scientific world view
This includes some very important issues like – it is possible to understand this highly organized world and its complexities with the help of latest technology. Scientific ideas are subject to change. After repeated experiments, a theory is established, but this theory could be changed or supported in future. Only laws that occur naturally do not change. Scientific knowledge may not be discarded but is modified – e.g., Albert Einstein didn't discard the Newtonian principles but modified them in his theory of relativity. Also science can't answer all our questions. We can't find answers to questions related to our beliefs, moral values and our norms.

Scientific inquiry
Scientific inquiry starts with a simple question. This simple question leads to information gathering, an educated guess otherwise known as hypothesis. To prove the hypothesis, an experiment has to be conducted, which yields data and the conclusion. All experiments must be repeated at least twice to get reliable results. Thus scientific inquiry leads to new knowledge or verifying established theories.
Science requires proof or evidence. Science is dependent on accuracy not bias or prejudice. In science, there is no place for preconceived ideas or premeditated results. By using their senses and modern technology, scientists will be able to get reliable information.
Science is a combination of logic and imagination. A scientist needs to think and imagine and be able to reason.
Science explains, reasons and predicts. These three are interwoven and are inseparable. While reasoning is absolutely important for science, there should be no bias or prejudice.
Science is not authoritarian, because history has shown that scientific authority has sometimes been proven wrong. No individual can determine or make decisions for others on any issue.

Scientific enterprise
Science is a complex activity involving various people and places. A scientist may work alone or in a laboratory, classroom or for that matter anywhere. Mostly it is a group activity requiring lot of social skills of cooperation, communication of results or findings, consultations, discussions etc.
Science demands a high degree of communication to the governments, funding authorities and to public.

e. Know that the ability of science and technology to resolve societal problems depends on the scientific literacy of a society

The most common definitions of science literacy are: scientific awareness (Devlin1998) and scientific ways of knowing (Maienshein 1999). In simple terms, scientific literacy is a combination of concepts, history, and philosophy that help us to understand the scientific issues of our time. The aim is to have a society that is aware of scientific developments.

The benefits for any society to be scientifically literate are –

1. To understand current issues
2. To appreciate the role of natural laws in ones life
3. To have an idea of the scientific advances

We are living in an age of scientific discoveries and technology. On TV and in the newspapers, we are constantly fed news related to science and technology. Scientific and technological issues are dominating our lives. We need to be scientifically literate to understand these issues. Understanding these debates has become as important as reading and writing. In order to appreciate the world around us and to be able to make informed personal decisions, we need to be scientifically literate.
It is the responsibility of the scientific community and educators to help the public to cope with the fast paced changes that are taking place now in the fields of science and technology.

Scientific literacy is based on the understanding of the most general principles and a broad knowledge of science. A society that is scientifically aware possesses facts and vocabulary sufficient to understand the context of the daily news. If one can understand articles about genetic engineering, the ozone hole, and greenhouse effect as well as sports, politics, arts, or the theater, then one is scientifically literate.

Scientific literacy is different from technological literacy and many times people are not clear about this. A survey indicated that less than 7% of the adults, 22% of college graduates and 26% of those with graduating degrees are scientifically literate. These numbers are not encouraging. In order to rectify this problem, more emphasis has been placed on science education in K-12 and at college level.

Skill 3.2   Diversity

a. Identify examples of women and men of various social and ethnic backgrounds with diverse interests, talents, qualities and motivations who are, or who have been, engaged in activities of science and related fields

Curiosity is the heart of science. Maybe this is why so many diverse people are drawn to it. In the area of zoology one of the most recognized scientists is Jane Goodall. Miss Goodall is known for her research with chimpanzees in Africa. Jane has spent many years abroad conducting long term studies of chimp interactions, and returns from Africa to lecture and provide information about Africa, the chimpanzees, and her institute located in Tanzania.

In the area of chemistry we recognize Dorothy Crowfoot Hodgkin. She studied at Oxford and won the Nobel Prize of Chemistry in 1964 for recognizing the shape of the vitamin B 12.

Have you ever heard of Florence Nightingale? She was a true person living in the 1800's and she shaped the nursing profession. Florence was born into wealth and shocked her family by choosing to study health reforms for the poor in lieu of attending the expected social events. Florence studied nursing in Paris and became involved in the Crimean war. The British lacked supplies and the secretary of war asked for Florence's assistance. She earned her nickname walking the floors at night checking on patients and writing letters to British officials demanding supplies.

In 1903 the Nobel Prize in Physics was jointly awarded to three individuals: Marie Curie, Pierre Curie, and Becquerel. Marie was the first woman ever to receive this prestigious award. In addition, she received the Nobel Prize in chemistry in 1911, making her the only person to receive two Nobel awards in science. Ironically, her cause of death in 1934 was of overexposure to radioactivity, the research for which she was so respected.

Neil Armstrong is an American icon. He will always be symbolically linked to our aeronautics program. This astronaut and naval aviator is known for being the first human to set foot on the Moon.

Sir Alexander Fleming was a pharmacologist from Scotland who isolated the antibiotic penicillin from a fungus in 1928. Flemming also noted that bacteria developed resistance whenever too little penicillin was used or when it was used for too short a period, a key problem we still face today.

Skill 3.3   Science, Technology, and Society

a. Identify and evaluate the impact of scientific advances on society

Society as a whole impacts biological research. The pressure from the majority of society has led to these bans and restrictions on human cloning research. Human cloning has been restricted in the United States and many other countries. The U.S. legislature has banned the use of federal funds for the development of human cloning techniques. Some individual states have banned human cloning regardless of where the funds originate.

The demand for genetically modified crops by society and industry has steadily increased over the years. Genetic engineering in the agricultural field has led to improved crops for human use and consumption. Crops are genetically modified for increased growth and insect resistance because of the demand for larger and greater quantities of produce.

With advances in biotechnology come those in society who oppose it. Ethical questions come into play when discussing animal and human research. Does it need to be done? What are the effects on humans and animals? There are no right or wrong answers to these questions. There are governmental agencies in place to regulate the use of humans and animals for research.

Science and technology are often referred to as a "double-edged sword". Although advances in medicine have greatly improved the quality and length of life, certain moral and ethical controversies have arisen. Unforeseen environmental problems may result from technological advances. Advances in science have led to an improved economy through biotechnology as applied to agriculture, yet it has put our health care system at risk and has caused the cost of medical care to skyrocket. Society depends on science, yet it is necessary that the public be scientifically literate and informed in order to prevent potentially unethical procedures from occurring. Especially vulnerable are the areas of genetic research and fertility. It is important for science teachers to stay abreast of current research and to involve students in critical thinking and ethics whenever possible.

b. Recognize that scientific advances may challenge individuals to reevaluate their personal beliefs

It is easy to say one is for or against something. Biotechnological advances are reaching new heights. This is both exciting and, to some, it creates anxiety. We are stretching our boundaries and rethinking old standards. Things we never thought possible, such as the human genome project, now seem ordinary, and cloning, once in the realm of science fiction, is now available. These revelations force us to rethink our stance on issues. It is normal to reevaluate one's beliefs. Reevaluation requires truly thinking about a topic, which in turn allows for recommitment to a topic or, possibly, a new, well thought out, position.

Skill 3.4   Safety

a. Choose appropriate safety equipment for a given activity (e.g., goggles, apron, vented hood)

It is the responsibility of the teacher to provide a safe environment for their students. Proper supervision greatly reduces the risk of injury and a teacher should never leave a class for any reason without providing alternate supervision. After an accident, two factors are considered; **foreseeability** and **negligence**. Foreseeability is the anticipation that an event may occur under certain circumstances. Negligence is the failure to exercise ordinary or reasonable care. Safety procedures should be a part of the science curriculum and a well managed classroom is important to avoid potential lawsuits. Students should wear safety goggles when performing dissections, heating, or while using acids and bases. Hair should always be tied back and objects should never be placed in the mouth. Food should not be consumed while in the laboratory. Hands should always be washed before and after laboratory experiments. In case of an accident, eye washes and showers should be used for eye contamination or a chemical spill that covers the student's body. Small chemical spills should only be contained and cleaned by the teacher. Kitty litter or a chemical spill kit should be used to clean spill. For large spills, the school administration and the local fire department should be notified. Biological spills should also be handled only by the teacher. Contamination with biological waste can be cleaned by using bleach when appropriate.
Accidents and injuries should always be reported to the school administration and local health facilities. The severity of the accident or injury will determine the course of action to pursue.

b. Discuss the safe use, storage, and disposal of commonly used chemicals and biological specimens

All laboratory solutions should be prepared as directed in the lab manual. Care should be taken to avoid contamination. All glassware should be rinsed thoroughly with distilled water before using and cleaned well after use. All solutions should be made with distilled water as tap water contains dissolved particles that may affect the results of an experiment. Unused solutions should be disposed of according to local disposal procedures.

The "Right to Know Law" covers science teachers who work with potentially hazardous chemicals. Briefly, the law states that employees must be informed of potentially toxic chemicals. An inventory must be made available if requested. The inventory must contain information about the hazards and properties of the chemicals. This inventory is to be checked against the "Substance List". Training must be provided on the safe handling and interpretation of the Material Safety Data Sheet.

The following chemicals are potential carcinogens and not allowed in school facilities: Acrylonitriel, Arsenic compounds, Asbestos, Bensidine, Benzene, Cadmium compounds, Chloroform, Chromium compounds, Ethylene oxide, Ortho-toluidine, Nickle powder, and Mercury.

Chemicals should not be stored on bench tops or heat sources. They should be stored in groups based on their reactivity with one another and in protective storage cabinets. All containers within the lab must be labeled. Suspect and known carcinogens must be labeled as such and segregated within trays to contain leaks and spills.

Chemical waste should be disposed of in properly labeled containers. Waste should be separated based on their reactivity with other chemicals.

Biological material should never be stored near food or water used for human consumption. All biological material should be appropriately labeled. All blood and body fluids should be put in a well-contained container with a secure lid to prevent leaking. All biological waste should be disposed of in biological hazardous waste bags.

Material safety data sheets are available for every chemical and biological substance. These are available directly from the company of acquisition or the internet. The manuals for equipment used in the lab should be read and understood before using them.

c. **Assess the safety conditions needed to maintain a science laboratory (e.g., eye wash, shower, fire extinguisher)**

All science labs should contain the following items of **safety equipment**. Those marked with an asterisk are requirements by state laws.

* fire blanket which is visible and accessible
*Ground Fault Circuit Interrupters (GCFI) within two feet of water supplies
*signs designating room exits
*emergency shower providing a continuous flow of water
*emergency eye wash station which can be activated by the foot or forearm
*eye protection for every student and a means of sanitizing equipment
*emergency exhaust fans providing ventilation to the outside of the building
*master cut-off switches for gas, electric and compressed air. Switches must have permanently attached handles. Cut-off switches must be clearly labeled.
*an ABC fire extinguisher
*storage cabinets for flammable materials
-chemical spill control kit
-fume hood with a motor which is spark proof
-protective laboratory aprons made of flame retardant material
-signs which will alert potential hazardous conditions
-containers for broken glassware, flammables, corrosives, and waste.
-containers should be labeled.

d. **Read and decode MSDS/OSHA (Material Safety Data Sheet/Occupational Safety and Health Administration) labels on laboratory supplies and equipment**

In addition to the safety laws set forth by the government regarding equipment necessary to the lab, OSHA (Occupational Safety and Health Administration) has helped to make environments safer by instituting signs that are bilingual. These signs use pictures rather than/in addition to words and feature eye-catching colors. Some of the best known examples are exit, restrooms, and handicap accessible.

Of particular importance to laboratories are diamond safety signs, prohibitive signs, and triangle danger signs. Each sign encloses a descriptive picture.

As a teacher, you should utilize a MSDS (Material Safety Data Sheet) whenever you are preparing an experiment. It is designed to provide people with the proper procedures for handling or working with a particular substance. MSDS's include information such as physical data (melting point, boiling point, etc.), toxicity, health effects, first aid, reactivity, storage, disposal, protective gear, and spill/leak procedures. These are particularly important if a spill or other accident occurs. You should review a few, available commonly online, and understand the listing procedures.

e. Discuss key issues in the disposal of hazardous materials in either the laboratory or the local community

Hazardous materials should never be disposed of in regular trash. Hazardous materials include many cleansers, paints, batteries, oil, and biohazardous products. Labels which caution one to wear gloves, to never place an item near another item (e.g., fire, electrical outlet), or to always use an item in a well ventilated area, should be taken as signals that the item is hazardous. Disposal of waste down the sink or in regular trash receptacles means that it will eventually enter the water/sewer system or ground, where it could cause contamination. Liquid remains/spills should be solidified using cat litter and then disposed of carefully. Sharps bins are used for the disposal of sharp objects and glass. Red biohazard bags/containers are used for the disposal of biohazard refuse.

f. Be familiar with standard safety procedures such as those outlined in the Science Safety Handbook for California Schools (1999)

Standard safety precautions include wearing gloves, using protective eye wear, and conducting experiments in appropriate areas (ventilated hoods when necessary) with appropriate equipment. Suggested safety guidelines are covered in depth in Part II, Domain 3, Section 3. In addition, the state of California has published a document entitled *Science Safety Handbook for California Schools.* This handbook can be purchased or printed at http://www.cde.ca.gov/pd/ca/sc/documents/scisafebk.pdf#search=%22CA%20science%20safety%20book%20for%20CA%20schools%22.

Sample Test

Directions: The following are multiple choice questions. Select from each grouping the best answer.

1. Which layer of the atmosphere would you expect most weather to occur?

   A. troposphere
   B. thermosphere
   C. mesosphere

2. What percentage of earth's surface is covered by water?

   A. 61%
   B. 71%
   C. 81%

3. Which layer of the earth's atmosphere contains the Ozone layer?

   A. thermosphere
   B. troposphere
   C. stratosphere

4. Copernicus developed a theory that is known as

   A. baycenter
   B. heliocentric
   C. geocentric

5. The boundary that separates the crust from the mantle is known as

   A. Moho
   B. shadow zone
   C. catacastic

6. The product of intrusive activities would result in forming a

   A. cinder cone
   B. volcanic pipe
   C. dike

7. A star's light and heat are produced by

   A. magnetism
   B. electricity
   C. nuclear fusion

8. The center of an atom is called

   A. micron
   B. nucleus
   C. electron

9. An important food source for animals in a water biome

   A. shrimp
   B. plankton
   C. seaweed

10. The smallest piece of an element is called a/an

    A. compound
    B. nucleus
    C. atom

11. An instrument that measures relative humidity is known as

    A. psychrometer
    B. anemometer
    C. barometer

12. Removing salts from ocean water by heating is called

    A. filtration
    B. distillation
    C. freezing

13. When molecules in the air cool and combine to form rain, _____ has occurred.

    A. condensation
    B. convection
    C. radiation

14. Which instrument measures wind direction?

    A. anemometer
    B. barometer
    C. wind vane

15. The most important cause of erosion is

    A. water
    B. wind
    C. air

16. An anemometer measures

    A. wind velocity
    B. temperature
    C. relative humidity

17. The boundary that develops when a cold air mass meets a warm air mass

    A. cold front
    B. warm front
    C. stationary front

18. When the sun, moon and earth are aligned in a straight line what type of tides are produced?

    A. neap tides
    B. high tides
    C. spring tides

19. North of the equator, currents move in which direction?

    A. counter-clock wise
    B. clockwise
    C. northerly

20. Rocks that serve as aquifers are

    A. impermeable
    B. permeable
    C. igneous

21. Volcanoes with violent eruptions are known as

    A. shield volcanoes
    B. dome volcanoes
    C. cinder volcanoes

22. The Richter scale measures

    A. compressions
    B. focus
    C. magnitude

23. The earth's outer core is probably

    A. liquid
    B. solid
    C. rock-bed

24. What are the two most abundant elements found in the earth's crust?

    A. oxygen and oxides
    B. oxygen and cabonates
    C. oxygen and silicon

25. The San Andreas Fault is classified as a

    A. transform fault
    B. oblique-slip fault
    C. reverse fault

26. Batholiths are the largest structures of which type of rock activity?

    A. intrusive rock
    B. extrusive rock
    C. magma

27. The main agents of chemical weathering are

    A. water, oxygen, CO_2
    B. water, oxygen, sulfur
    C. water, oxygen, nitrogen

28. Soil classified as porous is called

    A. clay soil
    B. laterites soil
    C. sandy soil

29. Intrusive igneous rock forms

    A. glassy texture
    B. small crystals
    C. large crystals

30. Which types of rocks are rich sources of fossil remains'?

    A. sedimentary rock
    B. metamorphic rock
    C. intrusive rock

31. The best preserved animal remains have been discovered in

    A. resin
    B. lava
    C. tar-pits

32. The Mid-Atlantic is a major area of which type plate movement?

    A. subduction plate movement
    B. divergent plate
    C. convergent plate

33. When lava cools quickly on the earth's surface the newly formed rock is called

    A. clastic
    B. intrusive
    C. extrusive

34. When a dyke forms with magma flowing in a tub-like structure this is known to be a/an

    A. extrusive activity
    B. intrusive activity
    C. metaphoric activity

EARTH & PLANETARY SCIENCE

35. A caldera is formed when a large depression collapses. This is the result of a

   A. sinkhole
   B. aquifer
   C. volcanic eruption

36. Trenches observed on the sea floor are the results of

   A. interaction
   B. divergence
   C. subduction

37. Alfred Wegener's hypothesis of continental drift was not supported until scientists began studying

   A. sea floor
   B. mountain ranges
   C. volcanoes

38. These massive waves are caused by the displacement of ocean water, and are often the result of underwater earthquakes.

   A. epicenters
   B. tidal waves
   C. tsunamis

39. Sea floor spreading occurs when the earth's crust is stretched and pulled apart in a process called

   A. slippage
   B. rifting
   C. drifting

40. The layer of the atmosphere that is in a plasma state and aids in communication

   A. Thermosphere
   B. Ionosphere
   C. Mesosphere

41. A stream erodes bedrock by grinding sand and rock fragments against each other. This process is defined as:

   A. dissolving
   B. transportation
   C. abrasion

42. Rocks formed from magma are:

   A. igneous
   B. metamorphic
   C. sedimentary

43. Rocks formed by the intense heating or compression of pre-existing rocks are classified as

   A. igneous
   B. metamorphic
   C. sedimentary

44. Rocks made of loose materials that have been cemented together are:

   A. igneous
   B. metamorphic
   C. sedimentary

45. River valley glaciers produce

   A. U-shaped erosion
   B. V-shaped erosion
   C. S-shaped erosion

46. The life cycle of a river with the most cutting power and erosion is known as which stage?

    A. youth stage
    B. mature stage
    C. old age

47. The result of radioactive decay

    A. parent element
    B. daughter element
    C. half-life

48. The most abundant dry gas found in the atmosphere is

    A. oxygen
    B. nitrogen
    C. CO_2

49. A natural groundwater outlet through which boiling water and steam explodes into the air is called a _____.

    A. sinkhole
    B. artesian system
    C. geyser

50. Which of the following rocks make the best aquifer?

    A. granite
    B. basalt
    C. sandstone

51. Sediments that settle out from rivers are called

    A. deposits
    B. boulders
    C. sandstone

52. A hole that remains in the ground after a block of glacier ice melts is called a

    A. pothole
    B. sinkhole
    C. kettle

53. The first sign that a tsunami is approaching a shore is

    A. a sudden flatten of the waves
    B. water moving from the shore
    C. large wall of water on horizon

54. Mountains that have been squeezed into wavelike patterns are called

    A. fold mountains
    B. dome mountains
    C. fault-block mountains

55. The largest ocean is the

    A. Atlantic
    B. Pacific
    C. Indian

56. The major surface current that flows along the east coast of the United States is known as the

    A. Bermuda Current
    B. Mexican Current
    C. Gulf Stream

57. The formation of ocean waves is caused by

    A. earth's rotation
    B. the moon
    C. the wind

58. The most abundant compound found in sea water is

    A. chloride
    B. calcium carbonate
    C. magnesium chloride

59. The distance between two meridians is measured in degrees of

    A. longitude
    B. latitude
    C. magnitude

60. A contour line that has tiny comb-like lines along the inner edge indicates a

    A. depression
    B. mountain
    C. valley

61. Fossils that are used to date strata are called

    A. datum fossils
    B. index fossils
    C. true fossils

62. Which of the following causes the aurora borealis?

    A. gases escaping from earth
    B. particles from the sun
    C. particles from the moon

63. The layer of the atmosphere that shields earth from harmful ultraviolet radiation is called

    A. ionic layer
    B. ozone layer
    C. equatorial layer

64. The layer of the earth's atmosphere that is closest to the earth's surface is the

    A. stratosphere layer
    B. thermosphere layer
    C. troposphere layer

65. The sun transfers its heat to other objects by

    A. conduction
    B. radiation
    C. convection

66. As an air mass expands, it becomes

    A. cooler
    B. warmer
    C. denser

67. Air moving northward from the horse latitudes produces a belt of winds called the

    A. prevailing westerlies
    B. north westerlies
    C. trade winds

68. Which type of cloud always produces precipitation?

    A. altostratus
    B. cirrostratus
    C. nimbostratus

69. An air mass that forms over the Gulf of Mexico is called

    A. polar
    B. maritime
    C. continental

70. Spring tides will occur when the moon is in its

    A. quarter phases
    B. full and new phases
    C. half phases

71. Air pressure is measured using a

    A. barometer
    B. hydrometer
    C. physcrometer

72. The two most abundant elements found in stars are

    A. hydrogen and calcium
    B. hydrogen and helium
    C. hydrogen and neon

73. A comet's tail always points _____ from the sun.

    A. towards
    B. perpendicular
    C. away

74. The dark areas observed on the sun are known as

    A. solar flares
    B. prominences
    C. sun spots

75. An example of distance in degrees of latitude is

    A. 55° north
    B. 93° east
    C. 25° west

76. A scale use to measure the hardness of a mineral is known as the

    A. Bowen's scale
    B. Mohs' scale
    C. Harding scale

77. When a gas changes to a liquid this process is known as

    A. evaporation
    B. condensation
    C. dissolution

78. A fan-shaped river deposit is better known as a

    A. levee
    B. flood plain
    C. delta

79. When heat energy is trapped by the gases in the Earth's atmosphere this process is called

    A. greenhouse effect
    B. coriolis effect
    C. constant effect

80. Winds in the Northern Hemisphere are deflected to the

    A. north
    B. left
    C. right

81. Water vapor and _____ trap heat in the atmosphere.

    A. carbon dioxide
    B. nitrogen
    C. sodium nitrate

82. The frontal system that forms when a cold air mass meets a warm air mass and does not change position is defined as a

    A. occluded front
    B. stationary front
    C. warm front

83. Surface ocean currents are caused by which of the following

    A. temperature
    B. density changes in water
    C. wind

84. The length of time it takes for two waves to pass in a row is called

    A. wave length
    B. wave period
    C. wave crest

85. Circulation of the deep ocean currents is the result of

    A. equatorial currents
    B. surface currents
    C. density currents

86. Chains of undersea mountains associated with the spreading of the seafloor are known as _____.

    A. ocean trenches
    B. mid ocean ridges
    C. seamounts

87. A shallow, calm area of water located between a barrier island and a beach area is defined as a/an _____.

    A. atoll
    B. coral reef
    C. lagoon

88. Closed contour lines noticed on a topographical map indicate which type of information?

    A. rivers and lakes
    B. hills
    C. mountains

89. The heliocentric model was developed by which famous scientist?

    A. Kepler
    B. Copernicus
    C. Newton

90. The phases of the moon are the result of its _____ in relation to the sun.

    A. revolution
    B. rotation
    C. position

91. A telescope that collects light by using a concave mirror and can produce small images is called a _____.

    A. radioactive telescope
    B. reflecting telescope
    C. refracting telescope

92. The measuring unit to measure the distance between stars is called

    A. astronomical unit
    B. light-year
    C. parsec

93. The largest planet found in the solar system is

    A. Pluto
    B. Jupiter
    C. Saturn

94. The famous scientist who discovered the elliptical orbits

    A. Kepler
    B. Copernicus
    C. Galilee

95. The planet with retrograde rotation is

    A. Pluto
    B. Uranus
    C. Venus

96. A star's brightness is referred to as

    A. magnitude
    B. mass
    C. apparent magnitude

97. Clouds of gas and dust where new stars originate are called

    A. black holes
    B. super novas
    C. nebulas

98. The transfer of heat from the earth's surface to the atmosphere is called ____.

    A. conduction
    B. radiation
    C. convection

99. The ozone layer is found in the

    A. stratosphere layer
    B. mesosphere layer
    C. exosphere layer

100. The coldest zone of the atmosphere is found in the

    A. thermosphere
    B. mesosphere
    C. stratosphere

101. Winds in high pressure areas tend to blow

    A. clockwise
    B. counterclockwise
    C. along the center

102. When warm air meets cold air this is defined as a

    A. cold front
    B. occluded front
    C. warm front

103. The fastest velocity of a river is found where?

    A. bottom
    B. center
    C. sides

104. As a glacier melts the sea level tends to:

    A. rise
    B. sink
    C. evaporate

105. The largest groups of minerals found in the earth's crust are

    A. silicates
    B. carbonates
    C. quartz

106. Used to measure the magnitude of an earthquake.

    A. Richter scale
    B. epicometer
    C. seismograph

107. These are types of folds:

    A. anticlines and synclines
    B. faults and folds
    C. fractures and shearings

108. Breaks in rocks which indicate movement are known as

    A. fractures
    B. folds
    C. faults

109. The collision of two continental plates is called a

    A. folded mountain range
    B. volcanic mountain range
    C. block mountain range

110. Plates that move in the same direction are termed

    A. divergent faults
    B. convergent faults
    C. transform faults

111. Studying the positions of layered rock is referred to as

    A. relative ages
    B. index fossils
    C. disconformity

112. The smallest division of geologic time is defined as

    A. Periods
    B. Eras
    C. Epochs

113. The most common fossils of the Paleozoic Era are

    A. angiosperms
    B. trilobites
    C. endotherms

114. Contamination may enter groundwater by

    A. air pollution
    B. leaking septic tanks
    C. photochemical processes

115. Which is a form of precipitation?

    A. snow
    B. frost
    C. fog

116. A dead star is called a _____.

   A. White Dwarf
   B. Super Giant
   C. Black Dwarf

117. Roughly ninety percent of all geologic time is said to be _____.

   A. Paleozoic
   B. Pre-Cambrian
   C. Mesozoic

118. The massive change in biological conditions that marked the beginning of life forms on earth is known as _____.

   A. Oxygen Revolution
   B. Carbon Revolution
   C. Trilobite Revolution

119. Water is a truly unique material. It has the property of _____.

   A. Adhesion
   B. Cohesion
   C. Both

120. The following is not a form of satellite used to track weather:

   A. NEXRAD
   B. Geostationary
   C. Polar Orbitting

121. Over the course of our planet's history Earth has had _____ atmosphere(s).

   A. one
   B. two
   C. three

122. Which is not a principle law of geology?

   A. Cross Cutting
   B. Faulting
   C. Super position

123. The red beds are important because they indicate the presence of _____ in the geologic record.

   A. Carbon
   B. Ammonia
   C. Oxygen

124. Tornadoes are most likely to occur in what season?

   A. Spring
   B. Summer
   C. Autumn

125. Which scale is used to measure hurricanes?

   A. Fujita Scale
   B. Saffir-Simpson Scale
   C. Richter Scale

Answer Key

1. A	45. A	89. B
2. B	46. A	90. C
3. C	47. B	91. B
4. B	48. B	92. C
5. A	49. C	93. B
6. C	50. C	94. A
7. C	51. A	95. C
8. B	52. C	96. A
9. B	53. B	97. C
10. C	54. A	98. A
11. A	55. B	99. A
12. B	56. C	100. B
13. A	57. C	101. A
14. C	58. A	102. C
15. A	59. A	103. B
16. A	60. A	104. A
17. A	61. B	105. A
18. C	62. A	106. C
19. B	63. B	107. A
20. B	64. C	108. C
21. C	65. B	109. A
22. C	66. B	110. C
23. A	67. A	111. A
24. C	68. C	112. C
25. A	69. B	113. B
26. A	70. B	114. B
27. A	71. A	115. A
28. C	72. B	116. C
29. C	73. C	117. B
30. A	74. C	118. A
31. C	75. A	119. C
32. B	76. B	120. A
33. C	77. B	121. C
34. B	78. C	122. B
35. C	79. A	123. C
36. C	80. C	124. A
37. A	81. A	125. B
38. C	82. B	
39. B	83. C	
40. B	84. B	
41. C	85. C	
42. A	86. B	
43. B	87. C	
44. C	88. B	

Rationales for Sample Questions

1. Which layer of the atmosphere would you expect most weather to occur?

   A. Troposphere

The troposphere is the lowest portion of the Earth's atmosphere. It contains the highest amount of water and aerosol. Because it touches the Earth's surface features, friction builds. For all of these reasons, weather is most likely to occur in the Troposphere.

2. What percentage of earth's surface is covered by water?

   B. 71%

The earth's surface is nearly ¾ covered with water. The Pacific ocean is the largest body of moving water. Of course there are other oceans, lakes, rivers, and glaciers as well.

3. Which layer of the earth's atmosphere contains the Ozone layer?

   C. Sratosphere

The stratosphere is located above the troposphere and below the mesosphere. It has layers striated by temperature. The warmest portion, the ozone layer, is warm because it absorbs solar ultraviolet radiation.

4. Copernicus developed a theory that is known as

   B. Heliocentric

Copernicus' theory stated that the planets revolved around the sun (helios), as opposed to prior belief that the planets revolved around the Earth (geocentric).

5. The boundary that separates the crust from the mantle is known as

   A. Moho

The Mohorovicic Discontinuity separates oceanic and/or continental crust from the Earth's mantle.

6. The product of intrusive activities would result in forming a

C. dike

A dike is formed when upwelling magma cools and solidifies beneath the surface, an intrusive activity.

7. A star's light and heat are produced by

C. nuclear fusion

Nuclear fusion is the process in which hydrogen atoms fuse together to form helium atoms, releasing massive amounts of energy during the fusion. It's the fusion of atoms, not combustion, which causes the star to shine.

8. The center of an atom is called

B. nucleus

The center of the atom is the nucleus. The nucleus of the atom is composed of nucleons, which when electrically charged are protons and when electrically neutral are neutrons. However, the electrons swirl around the nucleus in a large region called the Electron Cloud.

9. An important food source for animals in a water biome

B. plankton

Drifting organisms that inhabit the water column are called plankton. They may be phytoplankton or zooplankton. Phytoplankton are autotrophs and form the base of the aquatic food chain.

10. The smallest piece of an element is called a/an

C. atom

An atom is the smallest particle of the element that has the properties of that element. All of the atoms of a particular element are the same. The atoms of each element are different from the atoms of the other elements.

11. An instrument that measures relative humidity is known as

A. psychrometer

A psychrometer measures relative humidity. The other choices, anemometer and barometer, measure wind speed and atmospheric pressure, respectfully.

TEACHER CERTIFICATION STUDY GUIDE

12. Removing salts from ocean water by heating is called

    B. distillation

In the process of distilling ocean water the saline water is heated, producing water vapor that is in turn condensed, forming fresh water. The salt is left behind as waste but the water is used in many areas for drinking supply.

13. When molecules in the air cool and combine to form rain, _____ has occurred.

    A. condensation

Condensation is the change in matter from a denser phase, such as a gas (or vapor) to a liquid. Condensation commonly occurs when a vapor is cooled to a liquid.

14. Which instrument measures wind direction?

    C. wind vane

Of the choices given, an anemometer measures wind speed (velocity), a barometer measures atmospheric pressure and a wind vane indicates wind direction.

15. The most important cause of erosion is

    A. water

Erosion is most often caused by water. This can be acid rain eroding rocks, rivers eroding riverbeds, oceans eroding beaches and cliffs, etc. In addition, wind is another source of erosion.

16. An anemometer measures

    A. wind velocity

Of the choices given, an anemometer measures wind speed (velocity), temperature would be measured by a thermometer, and relative humidity is measured with a psychrometer.

TEACHER CERTIFICATION STUDY GUIDE

17. The boundary that develops when a cold air mass meets a warm air Mass

    A. cold front

Fronts are always labeled according to the approaching air mass. Therefore, a cold air mass meeting and displacing a warm air mass would be called a cold front.

18. When the sun, moon and earth are aligned in a straight line what type of tides are produced?

    C. spring tides

Spring tides are produced when the Earth, Sun, and Moon are in a line. Therefore, spring tides occur during the full moon and the new moon. Neap tides occur during quarter moons. They occur when the gravitational forces of the Moon and the Sun are perpendicular to one another (with respect to the Earth).

19. North of the equator, currents move in which direction?

    B. clockwise

North of the equator, currents move clockwise. South of the equator, currents move counter clockwise.

20. Rocks that serve as aquifers are

    B. permeable

Aquifers are underground areas of water-bearing permeable rock from which groundwater can be collected.

21. Volcanoes with violent eruptions are known as

    C. cinder volcanoes

Cinder volcanoes are some of the most violent volcanoes because of the immense pressure of gas built up within the neck of the volcanic tube. When it overcomes the resistance offered by the surrounding rock, it rips off the top of the cone. A huge mass of liquid magma and Pyroclastic Rock are flung outward in a violent explosion.

EARTH & PLANETARY SCIENCE

22. The Richter scale measures

   C. magnitude

The richter scale is used to measure the magnitude of earthquakes. Focus and compressions refer to areas of activity, but are not examples of a scale for measuring.

23. The earth's outer core is probably

   A. liquid

The earth's inner core is mathematically hypothesized to be a solid iron and nickel core. The outer core, surrounding the inner core, is so hot that it is believed to be molten iron (liquid state). Combined, they are responsible for Earth's magnetism.

24. What are the two most abundant elements found in the earth's crust?

   C. oxygen and silicon

Earth's crust is composed of 47% oxygen and 28% silicon.

25. The San Andreas Fault is classified as a

   A. transform fault

The San Andreas fault is considered a transform fault because sections of the earth's crust (the Pacific and North American Plates) slide side-by-side past each other.

26. Batholiths are the largest structures of which type of rock activity?

   A. intrusive rock

Batholiths are large portions of igneous intrusive rock deep within the Earth's crust that form from cooled magma.

TEACHER CERTIFICATION STUDY GUIDE

27. The main agents of chemical weathering are

    A. water, oxygen, CO_2

Water is the greatest factor in chemical weathering. Glaciers erode entire valleys. Rainfall pounds way at topographic surface features. Rivers erode riverbeds and river edges. Oceans erode shorelines and cliffs. Oxygen is also a factor in weathering. Air movement can have erosional factors. Most importantly, wind can transport material to other areas, having both erosional and depositional results. Carbon dioxide combines with water to produce carbonic acid, which erodes rock structures and some of our man made monuments.

28. Soil classified as porous is called

    C. sandy soil

Sandy soil has a high sand content. The sand molecules have many spaces in-between, making the soil porous. This soil does not hold water well.

29. Intrusive igneous rock forms

    C. large crystals

Intrusive igneous rock forms large crystals. This rock is formed from magma that cools and solidifies within the earth. Because it is surrounded by pre-existing rock, the magma cools slowly, and the rocks are coarse grained. The crystals are usually large enough to be seen by the unaided eye.

30. Which types of rocks are rich sources of fossil remains'?

    A. sedimentary rock

Sedimentary rock has the most abundant fossil collection. This is because, over time the layers of sand and mud at the bottom of lakes & oceans turned into rocks due to compression. Plants and animals that died and fell to the bottom were part of the compressional process by which the many layers were eventually turned into stone, encapsulating a fossil.

31. The best preserved animal remains have been discovered in

   C. tar-pits

Tar pits provide a wealth of information when it comes to fossils. Tar pits are oozing areas of asphalt, which were so sticky as to trap animals. These animals, without a way out, would die of starvation or be preyed upon. Their bones would remain in the tar pits, and be covered by the continued oozing of asphalt. Because the asphalt deposits were continuously added, the bones were not exposed to much weathering, and we have found some of the most complete and unchanged fossils from these areas, including mammoths and saber toothed cats.

32. The Mid-Atlantic is a major area of which type plate movement?

   B. divergent plate

The Mid- Atlantic is home to a submerged mountain range, which extends from the Arctic Ocean to beyond the southern tip of Africa. The divergent plate action results in sea floor spreading at a rate of about 2.5 centimeters per year (cm/yr), or 25 km in a million years, creating the vast ocean we recognize today.

33. When lava cools quickly on the earth's surface the newly formed rock is called

   C. extrusive

Rock formed by the cooling of magma on the earth's surface is known as extrusive, as opposed to intrusive, which is formed by the cooling of magma below the Earth's surface.

34. When a dyke forms with magma flowing in a tub-like structure this is known to be a/an

   B. intrusive activity

Dykes are thin, vertical veins of igneous rock. They form within fractures in the earth's crust. Intrusive activity forces magma into underground areas, which can seep into these existing fractures forming a dyke.

TEACHER CERTIFICATION STUDY GUIDE

35. A caldera is formed when a large depression collapses. This is the result of a

    C. volcanic eruption

A caldera is the collapse of land following a volcanic eruption. Once the underground store of magma and gas has been released in a volcanic explosion, there is not enough support, causing the ground to collapse. A caldera is sometimes confused with the area from which magma and gases are emitted (a crater).

36. Trenches observed on the sea floor are the results of

    C. subduction

Trenches are created where two plates collide (converge). Plate collision causes denser oceanic crust to sink or slip beneath lighter continental crust. It is subducted and melted into the asthenosphere, producing a deep trench on the ocean floor parallel to the plate boundary.

37. Alfred Wegener's hypothesis of continental drift was not supported until scientists began studying

    A. sea floor

Wegener's hypothesis of continental drift was supported by studies of the sea floor. In comparison to continental rock materials, the youngest rock is found on the ocean floor, consistent with the tectonic theory of cyclic spreading and subduction. Overall, oceanic material is roughly 200 million years old, while most continental material is significantly older, with age measured in billions of years.

38. These massive waves are caused by the displacement of ocean water, and are often the result of underwater earthquakes.

    C. tsunamis

Earthquakes can trigger an underwater landslide or cause sea floor displacements that in turn, generate deep, omni-directional waves. Far out to sea these waves may be hardly noticeable. However, as they near the shoreline, the shallowing of the sea floor forces the waves upward in a "springing" type of motion.

EARTH & PLANETARY SCIENCE

39. Sea floor spreading occurs when the earth's crust is stretched and pulled apart in a process called

B. rifting

A rift is a place where the Earth's crust and lithosphere are being pulled apart. In rifts, no crust or lithosphere is produced. If rifting continues, eventually a mid-ocean ridge may form.

40. The layer of the atmosphere that is in a plasma state and aids in communication

B. Ionosphere

The Ionosphere is an area of free ions: positively charged ions, produced as a result of solar radiation striking the atmosphere. It is known for its production of aurora borealis and its benefits to radio transmission

41. A stream erodes bedrock by grinding sand and rock fragments against each other. This process is defined as

C. abrasion

Abrasion is the key form of mechanical weathering. It is a sandblasting effect caused by particles of sand or sediment. Abrasive agents include wind blown sand, water movement, and the materials in landslides bashing into each other.

42. Rocks formed from magma are

A. igneous

Igneous rocks are rocks that have formed from cooled magma. They are further classified as extrusive or intrusive according to location.

43. Rocks formed by the intense heating or compression of pre- existing rocks are classified as

B. metamorphic

Metamorphism is the process of changing a pre-existing rock into a new rock by heat and or pressure. Metamorphism is similar to that of putting a clay pot into a kiln. The clay doesn't melt, but a solid-state chemical reaction occurs that causes a change. The chemical bonds of adjoining atoms breakdown and allow the atoms to rearrange themselves, producing a substance with new properties.

TEACHER CERTIFICATION STUDY GUIDE

44. Rocks made of loose materials that have been cemented together

    C. sedimentary

Sediments are broken up rock material. Sand on a beach or pebbles in a mountain stream are typical examples. Sedimentary rocks are named for their source; they are rocks that form from sediments that lithify to become solid rock. Sedimentary rock is especially important for the finding of fossils.

45. River valley glaciers produce

    A. U-shaped erosion

River valleys are typically v- shaped. The velocity and cutting power of a river is greatest at its center. However, glaciers broaden the area. Upon its retreat, a glacier typically leaves a U- shaped eroded valley.

46. The life cycle of a river with the most cutting power and erosion is known as which stage?

    A. youth stage

Young streams have straight paths, no flood plain, a "V" shaped cutting profile, and high velocity with generally clear water and low suspended load. Old streams have lots of meanders, large flood plain, flat profile, low velocity, with murky, "muddy" waters because of a high-suspended load.

47. The result of radioactive decay

    B. daughter element

The radioactive decay causes the (mother) element to change into an (daughter) element. The Mother-Daughter relationship of produced nuclides during the series of isotope decay is the basis for radiometric dating. Although many isotopes are used in radiometric dating, the most widely known method is referred to as Carbon-14 dating. Knowing the half-life (how long it takes for half of the material to decay) is the key factor in the radiometric dating process.

48. The most abundant dry gas found in the atmosphere is

    B. Nitrogen

The atmosphere is composed of 78% Nitrogen, 21% Oxygen, and 1% other gasses.

49. A natural groundwater outlet through which boiling water and steam explodes into the air is called a

　　C. geyser

A geyser is a thermal spring that erupts. The processes behind the eruption are very similar to those involved in boiling water in a teakettle. A constriction forms in the connected chambers of a spring. The water heats under pressure, turns to steam, and erupts with great force past the constriction. The ejected steam condenses and returns to a liquid state. The water draws back into its chambers and the process begins again. Since it takes awhile for the water to drain back and reheat, geysers often erupt on a determinable schedule.

50. Which of the following rocks make the best aquifer?

　　C. sandstone

Sandstone makes the best aquifer because of its porosity. It has larger pores than granite or basalt, and is also likely to fracture in a way that is condusive to water movement and collection.

51. Sediments that settle out from rivers are called

　　A. deposits

Deposits are pieces of matter that settle out of the water and fall to the bottom, or are washed into a collection area, such as a delta. This can be terrestrial matter, biological matter, salts, or larger pebbles and rocks.

52. A hole that remains in the ground after a block of glacier ice melts is called a

　　C. kettle

As the outwash moves sediment alongside and in the path of a receding glacier, blocks of ice can be buried beneath the sediment. After years of erosion these blocks are uncovered and melt, leaving a shallow depression behind. When these depressions fill, they are known as Kettles, and become scenic lakes.

TEACHER CERTIFICATION STUDY GUIDE

53. The first sign that a tsunami is approaching a shore is

    B. water moving from the shore

The first sign that a tsunami is approaching is usually the retreat of water from the shoreline. When the water returns, it comes fast and washes well past its normal level in both distance and depth, destroying coastal areas and causing many losses.

54. Mountains that have been squeezed into wavelike patterns are called

    A. fold mountains

During mountain building or compressional stress, rocks may deform to produce folds. Generally, a series is produced. The up-folds are called anticlines and the down-folds are known as synclines.

55. The largest ocean is the

    B. Pacific

The four major oceans (listed in decreasing size) are the Pacific, Atlantic, Indian and Arctic.

56. The major surface current that flows along the east coast of the United States is known as the

    C. Gulf Stream

The Gulf Stream begins in the Caribbean and ends in the northern North Atlantic. It is powerful enough to be seen from outer space and is one of the world's most studied current systems. It acts as the east coast boundary current plays an important role in the transfer of heat and salt to the poles.

57. The formation of ocean waves is caused by

    C. the wind

Wind is the primary factor in the production of ocean waves. It is the energy and friction of wind action that transfers to the water to create waves.

58. The most abundant compound found in sea water is

A. chloride

Chloride is the compound found most often in sea water. Other compounds commonly found include sodium carbonate, magnesium and potassium compounds, sulfite, bromide, and silicate. NaCl is what we commonly refer to as sea salt. Of the two components, chloride is more readily available in the sea.

59. The distance between two meridians is measured in degrees of

A. longitude

Longitude describes the location of a place on Earth east or west of a line called the Prime Meridian. Longitude is given in degrees ranging from 0° at the Prime Meridian to 180° east or west

60. A contour line that has tiny comb-like lines along the inner edge indicates a

A. depression

Contour lines are shown as closed circles in elevated areas and as lines with miniature perpendicular lined edges where depressions exist. These little lines are called hachure marks.

61. Fossils that are used to date strata are called

B. index fossils

Index fossils are fossils of organisms that were known to be abundant at specific times in Earth's history. Presence of such fossils gives one an idea of what age the surrounding material came from.

62. Which of the following causes the aurora borealis?

A. particles from the sun

Aurora Borealis is a phenomenon caused by particles escaping from the sun. The particles escaping from the sun include a mixture of gases, electrons and protons, and are sent out at a force that scientists call solar wind. Together, we have the Earth's magnetosphere and the solar wind squeezing the magnetosphere and charged particles everywhere in the field. When conditions are right, the build-up of pressure from the solar wind creates an electric voltage that pushes electrons into the ionosphere. Here they collide with gas atoms, causing them to release both light and more electrons.

63. The layer of the atmosphere that shields earth from harmful ultraviolet radiation is called

B. ozone layer

The ozone layer is the part of the Earth's atmosphere that contains high concentrations of ozone (O_3). It is located in the stratosphere and absorbs UV radiation emitted from the sun, making life possible on Earth.

64. The layer of the earth's atmosphere that is closest to the earth's surface is the

C. troposphere layer

The troposphere is the layer of Earth's atmosphere that is the lowest (closest to the surface). It is the densest because it contains almost all the water vapor and aerosol found in the atmosphere. It is easy to conclude, then, that most weather phenomena occur here.

65. The sun transfers its heat to other objects by

B. radiation

Radiation is the process by which energy is transferred in the form of waves or particles. The Sun emits ultraviolet radiation in UVA, UVB, and UVC forms, but because of the ozone layer, most of the ultraviolet radiation that reaches the Earth's surface is UVA.

66. As an air mass expands it becomes

B. warmer

Air expends as heat is applied according to the laws of gasses.

67. Air moving northward from the horse latitudes produces a belt of winds called the

A. prevailing westerlies

The prevailing westerlies are the winds found in the middle latitudes between 30 and 60 degrees latitude. They blow from the high pressure area in the horse latitudes towards the poles.

68. Which type of cloud always produces precipitation?

   C. nimbostratus

Nimbostratus clouds are seen as a thick, uniform, gray layer from which precipitation (significant rain or snow) is falling. Of the other choices offered, altostratus clouds appear as uniform white or bluish-gray layers that partially or totally obscure the sky, and cirrostratus are like a thin, nearly transparent, veil or sheet that partially or totally covers the sky. Only nimbostratus guarantees precipitation.

69. An air mass that forms over the Gulf of Mexico is called

   B. maritime

Maritime air masses are moist, containing considerable amounts of water vapor, which is ultimately condensed and released as rain or snow. Maritime tropical air originates near the Gulf of Mexico and travels north-east across the warm Atlantic to affect western Europe, as well as north-west across the United States.

70. Spring tides will occur when the moon is in its

   B. full and new phases

Spring tides are produced when the Earth, Sun, and Moon are in a line. Therefore, spring tides occur during the full moon and the new moon. Neap tides occur during quarter moons. They occur when the gravitational forces of the Moon and the Sun are perpendicular to one another (with respect to the Earth).

71. Air pressure is measured using a

   A. barometer

A psychrometer measures relative humidity. A barometer measures atmospheric pressure. A hydrometer is used to measure the specific gravity of a liquid.

72. The two most abundant elements found in stars are

   B. hydrogen and helium

Hydrogen and helium are the only elements that occur naturally in our universe. It makes sense, then, that they are present in all areas, including stars.

TEACHER CERTIFICATION STUDY GUIDE

73. A comet's tail always points _____ from the sun.

    C. away

A comet's tail always points away from the sun. The sun's radiation is burning up the ice that makes the comet, and since it is projecting the material outward, the tail seems to be pointing away from the sun. Notice that this question does not use a specific direction (north, south, east, west) because comets move and are subject to the viewer's location and perception.

74. The dark areas observed on the sun are known as

    C. sun spots

Larger dark spots called Sunspots appear regularly on the Sun's surface. These spots vary in size from small to 150,000 kilometers in diameter and may last from hours to months. The sunspots also cause solar flares that can accelerate to velocities of 900 km/hr, sending shock waves through the solar atmosphere.

75. An example of distance in degrees of latitude is

    A. 55° north

Latitude is measured in degrees away from the equator. The equator marks 0°, and parallel lines moving around the globe are quantified in degrees north or south.

76. A scale use to measure the hardness of a mineral is known as the

    B. Moh's scale

The Moh's scale of hardness measures the scratch resistance of minerals. The hardest material is diamond, and the frailest is talc. This means that diamond can scratch any surface, which is not true of less hard materials, such as talc.

77. When a gas changes to a liquid this process is known as

    B. condensation

Condensation is the change in matter to a denser phase, such as a gas (or vapor) to a liquid. Condensation can occur when a vapor is cooled to a liquid or when a vapor is compressed.

78. A fan-shaped river deposit is better known as a

  C. delta

Flowing water carries the material to the ocean where one of two things happen, the material is deposited on the offshore continental shelf or is carried back inland to the inlets and bays. Over time, the sediment thickly accumulates and may form typical coastal features such as sand bars and deltas.

79. When heat energy is trapped by the gases in the Earth's atmosphere this process is called

  A. greenhouse effect

When greenhouse gases and heat build up, the Earth's surface and atmospheric temperature rises. The current and controversial hypothesis contends that if we cut the amount of rising CO_2 in the atmosphere, then things will cool down.

80. Winds in the Northern Hemisphere are deflected to the

  C. right

The Earth is spinning on its rotational axis. Spin is greatest near the equator and least at the poles. The different velocities associated with the spin give rise to an effect on the air known as the Coriolis Force. The idea is that the result of the Coriolis effect is that winds in the north are deflected to the right, and winds in the south are deflected to the west.

81. Water vapor and _____ trap heat in the atmosphere.

  A. carbon dioxide

Water vapor and carbon dioxide are both considered greenhouse gases because they can trap heat in the atmosphere. Other sources of greenhouse gasses include rice paddies and ruminant animals, which produce Methane.

82. The frontal system that forms when a cold air mass meets a warm air mass and does not change position is defined as a

  B. stationary front

Fronts are the boundaries where one air mass meets another. A stationary front is a boundary between two air masses when neither is strong enough to displace the other.

83. Surface ocean currents are caused by which of the following

C. wind

A current is a large mass of continuously moving oceanic water. Surface ocean currents are mainly wind-driven and occur in all of the world's oceans (example: the Gulf Stream). This is in contrast to deep ocean currents which are driven by changes in density.

84. The length of time it takes for two waves to pass in a row is called

B. wave period

The wave period is the time required for two successive waves to pass. Wave crest is the tallest part of the wave. Wave length is measured from the crest of one wave to the crest of the next.

85. Circulation of the deep ocean currents is the result of

C. density currents

Unlike surface currents, deep ocean currents are driven by changes in density. These density differences may be caused by changes in salinity (halocline) or temperature (thermocline). Colder water sinks below warmer waters, causing a river (current) flowing below the warmer waters.

86. Chains of undersea mountains associated with the spreading of the seafloor are known as

B. mid ocean ridges

Mid ocean ranges are underwater mountains formed by plate tectonics. The underwater mountains are all connected, making a single mid-oceanic ridge system that is the longest mountain range in the world. The ridges are active sites with new magma constantly emerging onto the ocean floor and into the crust, resulting in sea floor spreading.

87. A shallow, calm area of water located between a barrier island and a beach area is defined as a/an

C. lagoon

A lagoon is known for its quiet movement of water. A lagoon is a body of shallow salt or brackish water separated from the sea by a shallow or exposed sandbank, coral reef, etc. Non-reef lagoon barriers are formed by wave-action or longshore currents depositing sediments. Because of their gentle atmosphere and brackish water, they are often nurseries for many baby fish and aquatic animals.

88. Closed contour lines noticed on a topographical map indicate which type of information?

B. hills

The rules of contouring dictate that contour lines are closed around hills, basins, or depressions. Because we know that depressions are shown using hachure marks, a closed contour line without such marks represents a hill.

89. The heliocentric model was developed by which famous scientist?

B. Copernicus

Copernicus is recognized for his heliocentric theory. The heliocentric theory postulates that the heavenly bodies rotate around the sun. Prior to his assertions, people believed in the geocentric model that held that all bodies rotated around the Earth. The geocentric model was supported by the church, so Copernicus' ideas were highly controversial.

90. The phases of the moon are the result of its _____ in relation to the sun.

C. position

The moon is visible in varying amounts during its orbit around the earth. One half of the moon's surface is always illuminated by the Sun (appears bright), but the amount observed can vary from full moon to none.

TEACHER CERTIFICATION STUDY GUIDE

91. A telescope that collects light by using a concave mirror and can produce small images is called a

    B. reflecting telescope

Reflecting telescopes are commonly used in laboratory settings. Images are produced via the reflection of waves off of a concave mirror. The larger the image produced the more likely it is to be imperfect.

92. The measuring unit to measure the distance between stars is called

    C. parsec

Parsecs are the units used to describe the distance between stars. Astronomical units (AU) are used to describe the distances between celestial objects (example The Earth is 1.00 ± 0.02 AU from the Sun). Light years are a unit of length measuring the distance light travels in a vacuum in one year.

93. The largest planet found in the solar system is

    B. Jupiter

The planets (in decreasing size) are Jupiter, Saturn (body- not inclusive of rings), Uranus, Neptune, Earth, Venus, Mars, Mercury (Pluto was thought to be the smallest planet, but is no longer classified as a planet).

94. The famous scientist who discovered the elliptical orbits

    A. Kepler

The significance of Kepler's Laws is that it overthrew the ancient concept of uniform circular motion, which was a major support for the geocentric arguments. Although Kepler postulated three laws of planetary motion, he was never able to explain *why* the planets move along their elliptical orbits, only that they did.

95. The planet with retrograde rotation is

    C. Venus

Venus has an axial tilt of only 3° and a very slow rotation. It spins in the direction opposite of its counterparts (who spin in the same direction as the Sun). Uranus is also tilted and orbits on its side. However, this is thought to be the consequence of an impact that left the previously prograde rotating planet tilted in such a manner.

96. A star's brightness is referred to as

A. magnitude

Magnitude is a measure of a star's brightness. The brighter the object appears, the lower the number value of its magnitude. The apparent magnitude is how bright an observer perceives the object to be. Mass has to do with how much matter can be measured, not brightness.

97. Clouds of gas and dust where new stars originate are called

C. nebulae

Nebulae are where new stars are born. They are large areas of gasses and dust. When the conditions are right, particles combine to form stars.

98. The transfer of heat from the earth's surface to the atmosphere is called

A. conduction

Radiation is the process of warming through rays or waves of energy, such as the Sun warms earth. The Earth returns heat to the atmosphere through conduction. This is the transfer of heat through matter, such that areas of greater heat move to areas of less heat in an attempt to balance temperature.

99. The ozone layer is found in the

A. stratosphere

The stratosphere is home to the ozone layer, which protects Earth from harmful UV radiation.

100. The coldest zone of the atmosphere is found in the

B. mesosphere

The mesosphere is the coldest layer of the atmosphere, with temperatures as low as -100°Celsius. Within this layer, temperature decreases with increasing altitude.

101. Winds in high pressure areas tend to blow

A. clockwise

High pressure systems are known for winds that flow clockwise and fair weather. Low pressure systems are accompanied by clouds and precipitation and winds flow counterclockwise.

102. When warm air meets cold air this is defined as a

C. warm front

When a warm air mass meets and displaces a cold air mass, the front is called a warm front.

103. The fastest velocity of a river is found where?

B. center

Mountain streams have little fining (sorting the material by size) due to their higher velocity, and low land streams are muddy because the velocity is less and erosion occurs on the bed and sides of the stream. Once a stream is at or close to base level, equilibrium is achieved between deposition and erosion. Erosion and deposition are controlled by the velocity of the stream. As the stream approaches base level, more of its energy is in a side-to-side cutting (meanders) than in down-cutting.

104. As a glacier melts the sea level tends to

A. rise

As a glacier melts, its water is distributed into nearby bodies of water, causing the sea level to rise.

105. The largest groups of minerals found in the earth's crust are

A. silicates

Silicates are the most abundant group of minerals found in the Earth's crust. The two most abundant elements in the earth's crust are Oxygen (46.6%) and Silicon (27.7%). These combine together to form silicates, which some scientists believe make up as much as 90% of the Earth's crust.

106. Used to measure the magnitude of an earthquake.

C. seismograph

A seismograph is a machine used to measure the magnitude of an earthquake. As the Earth's materials move, the weight also moves and sends an electronic signal to a recording device called a seismograph. Movements are displayed as a series of lines on a recording chart called a Seismogram, reflecting the seismic energy detected at a particular location.

107. These are types of folds:

A. anticlines and synclines

Folded mountains are composed of up and down folds. The up-folds are called anticlines and the down-folds are known as synclines.

108. Breaks in rocks which indicate movement are known as

C. faults

Faults are rock fractures that indicate relative movement. Fractures are also breaks in the rock, but they show no evidence of movement. Folds are created from compression and are a forming of tectonic building.

109. The collision of two continental plates is called a

A. folded mountain range

The collision of two continental plates results in a folded mountain range. Two continental plates pushing against each other but not subducting, will cause the material to buckle, sometimes repeatedly, giving these mountains their characteristic ribbon appearance.

110. Plates that move in the same direction are termed

C. transform faults

Transform faults are areas where two plates move in the same direction. They are parallel, and do not collide, but may result in earthquakes if areas of the plates stick or have excessive pressure in sliding past each other.

111. Studying the positions of layered rock is referred to as

A. relative ages

The Earth's materials-rocks, soils, and sediments-are piled upon each other in layers called strata. Understanding the relative orientation and arrangement of the strata provides important information about the Earth's history and the ongoing sequence of events and processes that helped shape that history.

112. The smallest division of geologic time is defined as

C. epochs

Geologic time is divided into eons, eras, periods, and epochs (listed here in decreasing order of size).

113. The most common fossils of the Paleozoic Era are

B. trilobites

Trilobites flourished in the Paleozoic era. There were over 600 genera and 1000's of species. Trilobites were bottom dwellers and scavengers found in shallow to deep water. For an extremely long period of time, Trilobites were the dominant multi-cellular life form on the planet. Trilobites are very good guide fossils because they were extremely abundant and existed throughout the entire Paleozoic period. Their development underwent distinctive changes, and these differences are useful in subdividing the time period.

114. Contamination may enter groundwater by

B. leaking septic tanks

Leaking septic tanks allow contamination to slowly seep into the ground, where it is absorbed into the water table and infects the groundwater.

115. Which is a form of precipitation?

A. snow

Snow is a form of precipitation. Precipitation is the product of the condensation of atmospheric water vapor that falls to the Earth's surface. It occurs when the atmosphere becomes saturated with water vapor and the water condenses and falls out of solution. Frost and fog do not qualify as precipitates.

TEACHER CERTIFICATION STUDY GUIDE

116. A dead star is called a _____.

    C. Black Dwarf

The final phase of a lower main sequence star's life cycle can take two paths: most main sequence white dwarfs after a few billion years completely burn out to become what is called a black dwarf: a cold, dead star. Alternatively, if a White Dwarf is part of a Binary Star: two suns in the same solar system, instead of slowly cooling to become a Black Dwarf, it may capture hydrogen from its companion star.

117. Roughly ninety percent of all geologic time is said to be _____.

    B. Pre-Cambrian

Pre-Cambrian Time: Comprised of the Hadean, Archean, and Proterozoic Eons, 87% of all geologic time is considered Pre-Cambrian.

118. The massive change in biological conditions that marked the beginning of life forms on earth is known as _____.

    A. Oxygen Revolution

Between 4.6 and 3.6 billion years ago, we transition from an uninhabitable Earth, to the appearance of simple, single-celled bacteria. Around 2.5 billion years ago, the bacteria developed the ability of photosynthesis. This process released oxygen as a by-product and there was a massive release of oxygen as the bacteria multiplied. This massive release is called the Oxygen Revolution and it concurrently marks the beginning of the Proterozoic Eon.

119. Water is a truly unique material. It has the property of _____.

    C. Both

A unique property of water is that water likes itself; it has a natural tendency to stick to itself. This property is based upon the polar nature of the water molecule. It attracts other water molecules. When the molecules stick together, they are attached through Hydrogen Bonds, giving the molecule a property called cohesion. Cohesion gives water an unusually strong surface tension, and its capillary action makes the water spread. When the water spreads, adhesion, the tendency of water to stick to other materials, allows water to adhere to solids, making them wet.

120. The following is not a form of satellite used to track weather:

   A. NEXRAD

While all of these instruments are used to track weather, the NEXRAD Radar, Next Generation Doppler Radar, is not a satellite. It emits beams of energy that are reflected by the water droplets in the atmosphere. This type of radar is very useful for tracking and predicting rain and less useful for snow or sleet. Geostationary satellites move with the Earth's rotation. Since they always look at the same point, this allows for a view showing changes over periods of time. Polar Orbiting satellites follow an orbit from pole to pole. The Earth rotates underneath the satellite and gives a view of different areas. In effect, it produces slices of the Earth.

121. Over the course of our planet's history Earth has had _____ atmosphere(s).

   C. three

Earth's initial atmosphere was composed of primarily hydrogen and smaller amounts of helium. However, most of the hydrogen and helium escaped into space very shortly after the earth was formed, approximately 4.6 billion years ago. A second atmosphere formed during the first 500 million years of Earth's history, as the gasses trapped within the planet were out-gassed during volcanic eruptions. This atmosphere was composed of carbon dioxide (CO_2), Nitrogen (N), and water vapor (H_2O), with smaller amounts of methane (CH_4), ammonia (NH_3), hydrogen (H), and carbon monoxide (CO). However, only trace quantities of oxygen were present. At around 3.5 billion years, Earth's third atmosphere began to form as the first life forms- simple, unicellular bacteria- appeared.

122. Which is not a principle law of geology?

   B. faulting

The principle laws of geology are:
- Principle of Uniformitarianism: Processes that are happening today also happened in the past.
- Principle of Cross-Cutting Relations: A rock is younger than any rock it cuts across.
- Principle of Original Horizontality: Rock units are originally laid down flat. Something happened to cause them to change orientation.
- Principle of Super Position: The rock on the bottom is older than the rock on top.
- Principle of Biologic Succession: Fossils correspond to particular periods of time.

123. The red beds are important because they indicate the presence of _____ in the geologic record.

    C. Oxygen

Formation of Red Beds: The Animike Group- banded iron formations- form. These Red Beds are important because they herald the appearance of significant amounts of oxygen on the Earth. The red color is produced by rust. The rust indicates the presence of oxygen acting upon the ferrous material present in the ocean, and eventually, on the land. The presence of significant amounts of oxygen allows ozone to form, which in turn, screens out the harmful ultra-violet (UV) rays. This makes life possible outside of the protective confines of the ocean.

124. Tornadoes are most likely to occur in what season?

    A. Spring

Tornado: an area of extreme low pressure, with rapidly rotating winds beneath a cumulonimbus cloud. Tornadoes are normally spawned from a Super Cell Thunderstorm. They can occur when very cold air and very warm air meet, usually in the Spring. Tornadoes represent the lowest pressure points on the Earth and move across the landscape at an average speed of 30 mph.

125. Which scale is used to measure hurricanes?

    B. Saffir-Simpson Scale

The Fujita Scale is used to measure the intensity and damage associated with tornadoes. The Saffir-Simpson Scale is used to classify hurricanes into five categories, with increasing numbers corresponding to lower central pressures, greater wind speeds, and large storm surges. Richter Scale: the primary scale used by seismologists to measure the magnitude of the energy released in an earthquake.

XAMonline, INC. 21 Orient Ave. Melrose, MA 02176
Toll Free number 800-509-4128
TO ORDER Fax 781-662-9268 OR www.XAMonline.com

<u>CALIFORNIA SUBJECT EXAMINATIONS   - CSET - 2008</u>

PO#   Store/School:

Address 1:

Address 2 (Ship to other):

City, State Zip

Credit card number_____-_____-_____-_____   expiration_____

EMAIL _____

PHONE   FAX

ISBN	TITLE	Qty	Retail	Total
978-1-58197-595-6	RICA Reading Instruction Competence Assessment			
978-1-58197-596-3	CBEST CA Basic Educational Skills			
978-1-58197-398-3	CSET French Sample Test 149, 150			
978-1-58197-622-9	CSET Spanish 145, 146, 147			
978-1-58197-803-2	CSET MSAT Multiple Subject 101, 102, 103			
978-1-58197-261-0	CSET English 105, 106, 107			
978-1-58197-608-3	CSET Foundational-Level Mathematics 110, 111			
978-1-58197-285-6	CSET Mathematics 110, 111, 112			
978-1-58197-340-2	CSET Social Science 114, 115			
978-1-58197-342-6	CSET General Science 118, 119			
978-1-58197-585-7	CSET Biology-Life Science 120, 124			
978-1-58197-395-2	CSET Chemistry 121, 125			
978-1-58197-399-0	CSET Earth and Planetary Science 122, 126			
978-1-58197-224-5	CSET Physics 123, 127			
978-1-58197-299-3	CSET Physical Education, 129, 130, 131			
978-1-58197-397-6	CSET Art Sample Subtest 140			
			SUBTOTAL	
			Ship	$8.70
			TOTAL	

www.ingramcontent.com/pod-product-compliance
Lightning Source LLC
Chambersburg PA
CBHW080540300426
44111CB00017B/2816